高等学校教材

科技英语翻译教程

主　编　赵玉闪　廖　麦

副主编　高晓薇　刘　阳　张　帆

　　　　金　英　吕亮球　刘　军

参　编　冯俊宝　张　湛　郭晓军

　　　　尹　宇　谷杏芬　杨树国

　　　　史　娟　董时雨　岳　贞

中国计量出版社

图书在版编目（CIP）数据

科技英语翻译教程/赵玉闪，廖麦主编．—北京：中国计量出版社，2011.3（2023.8 重印）
高等学校教材
ISBN 978－7－5026－3412－4

Ⅰ.①科…　Ⅱ.①赵…②廖…　Ⅲ.①科学技术—英语—翻译—高等学校—教材
Ⅳ.①H315.9

中国版本图书馆 CIP 数据核字（2011）第 016899 号

内　容　提　要

本书对科技英语的翻译方法及翻译技巧进行了详细的分析讲解。书中配有大量的翻译实例，力求将翻译理论与翻译实践有机结合。每一章都配有相应的翻译练习题，并给出了参考答案。

本书适用于非英语专业研究生的翻译课教学，同时也适合不同领域的科研工作者、从事翻译教学和实践的人士参考使用。

中国计量出版社出版
北京和平里西街甲 2 号
邮政编码　100013
电话（010）64275360
http://www.zgjl.com.cn
中国标准出版社秦皇岛印刷厂印刷
新华书店北京发行所发行

*
787 mm×1092 mm　16 开本　印张 12.5　字数 295 千字
2011 年 3 月第 1 版　　2023 年 8 月第 4 次印刷
*
定价：26.00 元

前　言

随着我国经济的发展、科学技术的进步和对外信息交流的日益增多，科技文献资料的翻译日益为人们所重视。科技英语作为英语的一种语体，在词汇、语法、修辞等方面均有别于其他英语语体，具有自己鲜明的特色。为了更好地促进我国的科技进步，熟悉科技领域研究的国内外动态、发展方向，广泛开展国际交流与合作，学习和掌握科技英语的翻译方法和技巧，不仅是专业翻译工作者，也是科研技术人员的一门必修课。

非英语专业研究生通常把英语作为专业研究的辅助工具。研究生进入专业研究领域后经常要进行与本专业相关的文献资料、学术论文、学术著作、研究报告、专利产品与说明书等英文科技文体的翻译工作。由于科技英语注重科学性、逻辑性、正确性与严密性，科技文体的翻译要求用词准确、规范、语气正式、陈述客观、逻辑性强。然而，未接受过翻译系统训练的学生在翻译实践中普遍存在感性翻译的盲目性，仅凭语言直觉反映能力进行翻译，很少主动运用翻译技巧，从而严重影响了翻译质量和效率。因此，本书的编写，既是为了具体贯彻和实施教育部教学大纲中针对学生翻译能力训练和培养的指导方针，也是为了满足研究生公共英语教学实践的迫切需要。

本书的设计和编写注重翻译理论与翻译实践的有机结合。为了帮助学生卓有成效地提高翻译技能，本书通过大量翻译实例详细讲解了翻译的基础知识、翻译标准、翻译方法和技巧，并在每一章后都附有相应的翻译练习，以巩固有关理论知识，培养熟练的翻译技能。

本书从内容安排上遵循深入浅出、由点到面的思路。首先，介绍了科技英语的语言特点、翻译方法、标准、过程和技巧。其次，逐级分析了从词、短语到句子的英汉对比及其译法。再次，点面结合地说明了科技英语在语态、时态和语气上的翻译以及科技文章段落的翻译。最后，强调了标点符号的翻译在追求准确、严谨的科技英语翻译中的重要性。

本书主要适用于非英语专业研究生的翻译课教学，同时也适合不同领域的科技工作者、从事翻译教学和实践的人士参考使用。

本书由华北电力大学、北京物资学院、北方工业大学、内蒙古工业大学、河北大学等院校的老师共同编写。由于编者水平有限、时间仓促，疏漏之处在所难免，希望广大读者在使用中提出意见和建议，使本书更臻完善。

编　者

2011 年 1 月

目　录

第1章　科技英语的语言特征

随着全球经济一体化的逐步深入，现代科学技术的飞速发展，在日益广泛的国际科学技术交流和合作中，科技英语（English for Science and Technology，EST）越来越引起科学界和语言界的高度重视和关注。

科技文体严谨周密、概念准确、逻辑性强、行文简练、重点突出、句式严整、少有变化，常使用前置性陈述，即在句中将主要信息尽量前置，通过主语传递主要信息。我国著名科学家钱三强教授曾指出，科技英语无论在语法结构还是词汇方面都逐渐形成了特有的习惯用法、特点和规律。这就是说，科技英语除了具有与普通英语相同的共性之外，又具有个性。科技英语的语言特征主要体现在词汇、句法和修辞三个方面。

1.1　词汇特征

科技词汇多源于希腊语和拉丁语，这两种语言是世界上成熟最早和最完备的语言，词汇无词形、词义上的变化，具有稳定性。在科技英语里大量使用的词汇主要包括纯科技词汇、通用科技词汇和派生词汇三大类。

1.1.1　纯科技词汇

纯科技词汇指的是只用于某个专业或学科的专门词汇或术语。相对而言，专业词汇词义专一、含义明确，一词多义的情况较少。随着科技的发展和新学科、新专业的产生，这类词汇层出不穷。这些专业术语的特点是词形较长，大多含有源于拉丁语、希腊语和法语的词根、词缀。这类词语的语义范围较为狭窄，针对性极强，意义较为明确固定，符合科技英语准确明晰的要求。如：

nucleonics　核子学
semisomnus　半昏迷
isotope　同位素
cryogenics　低温学
promethazine　异丙嗪
plancton　浮游生物
hydroxide　氢氧化物
diode　二极管
norepinephrine　降肾上腺素
autoradiography　自动射线照相术
excoriation　表皮脱落

1.1.2　通用科技词汇

通用科技词汇指的是不同专业都要经常使用的那些词汇，数量较大。这类词的使用范围比纯科技词汇要广，出现频率也高，但在不同的专业里有较为稳定的词义。如：

frame　框架（一般意义）；机架（机械学）；帧、镜头（电讯）
normal　正常的（一般意义）；中性的、当量的、标准浓度（化学）；简正的（物理

学)；垂直的、法线的、法线（数学)；不受感染的（生理学）

transmission 发射（无线电工程学)；传动、变速（机械学)；透射（物理学)；遗传（医学）

power 乘方、次方、幂（数学)；力、电、电力、动力、电源、功率（机械力学）

adjustment 调整（会计学)；海损理算（保险）

award 裁决（仲裁)；决算（招标）

performance 履行（贸易)；性能（机械）

operation 操作（工程)；手术（医学)；经营、管理（管理学）

以 run 为例：

例 1 The generator has been running for days.

【译文】那台发电机已经运转了好几天。

例 2 The car is running at 80 miles an hour.

【译文】汽车正以每小时 80 英里的速度行驶。

例 3 He ran the molten metal into a mould.

【译文】他把融化的金属倒进模具里。

例 4 The motor runs up quickly to the normal speed.

【译文】电动机很快达到正常速度。

例 5 The battery has run down.

【译文】电池耗尽了。

1.1.3 派生词汇

派生词汇指通过合成、转化和派生构词手段而形成的词汇，表示科技发展中出现的新事物。这种词汇在科技英语文献中占有很大的比重。据统计，由前缀 anti-、non-、hydro-、hyper-、hypo-、inter- 与不同的词根构成的词条在科技英语中就有两千多个。以表示学科的后缀 -logy、-ics 和表示行为、性质、状态等的后缀 -tion、-sion、-ance、-ence、-ment 构成的词汇在科技英语文献中也很多。新词出现时，要懂得利用构词法理解词义并准确恰当地译出。如：

antiphase 反相

nonconducting 不传导的

hydrodynamics 流体动力学

insulation 绝缘

waterlock 水闸

sleeping-pill 安眠药

moonwalk 月球漫步

skylab 太空实验室

minicell 微细胞

minicrystal 微晶体

space junk 太空垃圾

radiophotography 无线电传真

anti-armored fighting vehicle missile 反装甲车导弹

friction factor 摩擦系数

thermocouple 热电偶

voltmeter 电压表

另外，为了使用便利和节省时间，科技英语同其他文体英语一样，也有许多缩略词。如：

IC（integrated circuit） 集成电路

DNA（deoxyribonucleic acid）　脱氧核糖核酸
CCRR（co-channel rejection ratio）　同频抑制比
cpd（compound）　化合物
FM（frequency modulation）　调频
telesat（telecommunications satellite）　通讯卫星
HSCDS（high speed circuit switched data）　高速电路交换数据
HPCC（high performance computing and communication program）　高性能计算机与通信规划

由动词或名词派生出来的形容词作描绘性词语（descriptive word）的比较多，特别是用来表示数量、大小、程度、性质、状态、形状等意义。常用的形容词后缀有：-ac/-iac、-al、-ar、-ary、-ato、-eal、-ed、-ic、-ible/-able、-ing、-ive、-oid、-ose、-ous、-y等。如：

preventive measure　预防措施
basic dyes　碱性染料
speedy reply　迅捷答复
tubular organ　管状器官
muscular activity　肌肉活动
involved tissue　受累组织
contagious disease　接触性传染病
bacterial infection　细菌感染
favorable prognosis　预后良好
systemic disorder　全身紊乱

实践中，科技英语中最难理解和最难翻译的不是专业词汇，而是一些含有科技意义的动词、副词和形容词，特别是一些短语动词。这要求译者了解多义词的每一个含义及在文章中的语境，并能通过适当的专业知识加以判断来选择词义。

1.2　句法特征

1.2.1　被动语态

在各种文体中，英语的被动语态要比汉语中使用得多，在科技英语中尤为突出。科技英语的宗旨是要阐述客观事物的本质特征，描述其发生、发展及变化过程，表述客观事物间的联系，所以它的主体通常是客观事物或自然现象，这样一来，被动语态也就得以大量使用。此外，被动语态所带有的叙述客观性也使得作者的论述更显科学性，从而避免主观色彩。与这一特点相适应的是科技英语中少用第一人称和第二人称，即便是非用不可也常常是使用它们的复数形式以增强论述的客观性。

从语言结构上来讲，科技英语许多程式化的句子出现频率较高，翻译过程中应该熟悉其惯译方法。如：

It cannot be denied that...　不可否认……
It can be seen that...　可以看出……
It is generally/accepted/recognized/regarded/that...　普遍认为（一般认为或大家公认）……
It is estimated that...　据估计，据推算……

类似的句型还有：

be accepted as... 被承认为……
be accounted as... 被认为是……
be adopted as... 被用作为……
be classified as... 被划分为……
be conceived of as... 被想象为……
be considered as... 被认为是……
be counted as... 被当作……
be described as... 被描述为……
be expressed as... 被表示为……
be employed as... 被用作……
be known as... 被认为是……

例 6 New computer viruses and logic bombs are discovered every week.

【译文】每个星期，我们都能发现新型计算机病毒和逻辑炸弹。

例 7 Computers can also be used to automate shipbuilding. That has already been done in many ways. Not long ago steel plates were still marked out by hand. This method was replaced by optical following. Therefore steel plates are cut automatically by gas burners. Nowadays the cutting machine is controlled by information stored on computer tape. This is called numerical control.

【译文】计算机还可用于使造船业实现自动化，这在许多方面已经实现了。不久前钢板还要用手工画出大样，后来光学跟踪代替了这一方法。因而钢板能用气割枪自动切割出来。现在则是用计算机磁带中储存的信息控制切割机，这就叫做数字控制方法。

例 8 When the radiant energy of the sun falls on the earth, it is changed into heat energy, and the earth is warmed.

【译文】太阳的辐射能到达地球后就转化为热能，从而使大地暖和起来。

例 9 The Harry Diamond Laboratories performed early advanced development of the Arming Safety Device (ASD) for the Navy's 5-in. guided projectile. The early advanced development was performed in two phases. In phase 1, the ASD was designed, and three prototypes were fabricated and tested in the laboratory. In phase 2, the design was refined, 35 ASDs and a large number of explosive mockups were fabricated, and a series of qualification tests was performed. The qualification tests ranged from laboratory tests to drop tests and gun firing. The design was further refined during and following the qualification tests. The feasibility of the design was demonstrated.

【译文】哈里·代蒙德实验室对美海军 5 英寸制导炮弹的解除保险装置（ASD）进行了预研。预研工作分两个阶段进行。第一阶段，先设计出 ASD，并试制三个样件在实验室进行试验；第二阶段，对原设计进行改进并制造出 35 个 ASD 和大量的爆炸模型，接着进行了一系列鉴定试验。试验包括实验室试验、落锤试验和火炮射击试验。在试验期间和试验结束后，又对设计做了进一步的改进。设计方案的可行性已经得到证明。

1.2.2 非谓语动词

在科技英语中大量使用非谓语动词，可以更好、更准确地描述各事物之间的关系，事物的位置和状态的变化。使用非谓语动词使句子结构严谨、逻辑性强。如：

例 10 Cement, wood and steel are the most widely used building materials.

【译文】水泥、木材和钢材是最广泛使用的建筑材料。

例 11 In communications, the problem of electronics is how to convey information from one place to another.

【译文】在通讯系统中，电子学要解决的问题是如何把信息从一个地方传递到另一个地方。

例12 A safety valve is provided in order to allow excess pressure to escape.

【译文】安全阀的设置能降低过高的压力。

例13 The volume of fuel oil extracted from the liquid produced increases substantially.

【译文】从所产生液体中提取的燃料油的数量大大增加。

例14 China was the first country to invent rockets.

【译文】中国是第一个发明火箭的国家。

例15 Astronauts performing tasks outside of space shuttle get help from robot arm.

【译文】宇航员在航天飞机外面执行任务时可借助于机械手。

1.2.3 名词化结构

用表示动作或状态的抽象名词词组或短语（主要是用具有动作意义的名词 + of + 修饰语）来表示一个句子的意思，就是名词化结构。这一结构具有简洁、准确、严密、客观、信息量大等特点，在科技英语中大量使用。如：

the generation of heat by friction 摩擦生热

the transmission and reception of images of moving objects by radio waves 通过无线电波来发射和接收活动物体的图像

computer programming teaching device manual 计算机程序编制教学装置手册

例16 We must place stress on the prevention of diseases.

【译文】我们应以预防疾病为主。

例17 The construction of such bridges has now been realized; its realization being supported with all the achievements of modern science.

【译文】这种桥梁的结构现在已经实现，它的实现是现代科学的伟大成就。

1.2.4 长句和逻辑关联词

科技英语中虽然大量使用名词化词语、名词短语结构以及悬垂结构来压缩句子长度，但是为了将事理充分说明，也常常使用一些含有许多短语和分句的长句，同时还常使用许多逻辑性语法词，如表示原因的 because（of）、due（owing）to、as（a result of）、caused by、for、since、now that、in that 等，表示语气转折的 but、however、nevertheless、otherwise、yet 等，表示逻辑顺序的 so、therefore、thus、furthermore、moreover、in addition to 等，以使行文逻辑关系清楚，层次条理分明。

在科技英语翻译中要有意识地识别句子中的形态词（名词、动词、形容词和副词等，这些词有一定形态符号特征）和结构词（介词、连词、冠词、关系代词和关系副词等），了解句子的基本句型、成分和语法关系，进而深入了解句子语言成分的概念范畴之间的关系，即主题关系。也就是说，在遇到长句时，需要通过形态识别，突显主、谓、宾、表等主干成分，了解其“骨架含义”及次要成分的含义，理解这些成分之间的逻辑关系和修饰关系，然后通过适当的方法翻译出来。

例18 A gas may be defined as a substance which remains homogeneous, and the volume of which increases without limit, when the pressure on it continuously reduced, the temperature being maintained constant.

【译文】气体是一种始终处于均匀状态的物质，当温度保持不变，而其所受的压力不断降低时，它的体积可以无限增大。

原句虽然不长，但包含两个由 which 引导的定语从句，一个由 when 引导的状语从句。

例 19 One of the most important things which the economic theories can contribute to the management science is building analytical models which help in recognizing the structure of managerial problem, eliminating the minor details which might obstruct decision-making, and concentrating on the main issues.

【译文】经济理论对于管理科学的最重要贡献之一，就是分析模型的建立，这种模型有助于认识管理问题的构成，排除可能妨碍决策的次要因素，从而有助于集中精力去解决主要的问题。

1.2.5 介词词组

为了较为简练地反映各事物、各句子成分之间的时空、所属、因果等逻辑关系，科技英语中介词词组、短语使用较多。

例 20 A body in motion remains in motion unless acted on by an external force.

【译文】如果没有外力的作用，运动的物体仍然保持运动状态。

例 21 The action of air on an airplane in flight at low altitude is greater than that at high altitude.

【译文】空气对于低空飞行飞机的作用力大于高空飞行的飞机。

1.3 修辞特征

1.3.1 时态

从时态的角度看，科技英语中多使用一般现在时和完成时。这是因为一般现在时可以较好地表现文章内容的无时间性，说明文章中的科学定义、定理、公式、现象和过程等不受时间限制，任何时候都成立；而完成时则多用来表述已经发现或获得的研究成果。提到别人的发现或报道时常用 Those authors have found that..., Someone has reported that... 等句型。

例 22 The general layout of the illumination system and lenses of the electron microscope corresponds to the layout of the light microscope. The electron "gun" which produces the electrons is equivalent to the light source of the optical microscope. The electrons are accelerated by a high-voltage potential (usually 40000 to 100000 volts), and pass through a condenser lens system usually composed of two magnetic lenses. The system concentrates the beam on to the specimen, and the objective lens provides the primary magnification. The final images in the electron microscope must be projected on to a phosphor-coated screen so that it can be seen. For this reason, the lenses that are equivalent to the eyepiece in an optical microscope are called "projector" lenses.

【译文】电子显微镜的聚光系统和透镜的设计与光学显微镜的设计是一致的。电子

“枪”可以产生电子束，电子束相当于光学显微镜的光源。电子被高压（通常为40000~100000伏）的电位差加速，穿过聚光镜系统。聚光镜通常由两组磁透镜组成。聚光镜系统可将电子束聚集在样品上，并且物镜可对样品进行初级放大。电子显微镜的最终成像被投射到涂磷的荧光屏上，以便进行观察。正是由于这个原因，这些相当于光学显微镜目镜的透镜被称为“投影镜”。

例23　Humans have been dreaming of copies of themselves for thousands of years.

【译文】千百年来，人类一直梦想制造出自己的复制品。

例24　By the turn of the 19th century geologist had found that rock layers occurred in a definite order.

【译文】到19世纪初，地质学家已发现岩层有一定的次序。

1.3.2　虚拟语气的使用

科技作者在说明事理、提出设想、探讨问题和推导公式时，常常涉及各种前提、条件和场合，为了避免武断，总是从假定、猜测、建议和怀疑的角度出发，这就往往需要采用虚拟语气；另一方面，有不少作者为了表示自己的谦逊，也经常采用虚拟语气以使口吻变得委婉。

例25　If it moves，an electron produces a magnetic field.

【译文】如果电子运动的话，它就会产生磁场。

如果if从句中提出的条件根本不可能实现或者至少说话者认为不可能实现时，即所提出的假设与过去、现在或将来的事实相反时，那就是虚拟条件句。

例26　If the dam had broken，we would have been killed.

【译文】如果大坝决口的话，我们早就死了。

例27　If there were no gravity，there would be no air round the earth.

【译文】倘若没有重力，地球周围就不会有空气。

例28　If she were to do this test，she might do it in some other way.

【译文】要是她来做这项实验，她可能用另外一种方法去做。

例29　If a pound of sand were broken up and turned into atomic energy，there would be enough power to supply the whole of Europe for a few years.

【译文】如果使一磅沙子分裂并转变成原子能，那么这种能量将足以供整个欧洲使用好几年。

例30　Reduce the sun to the size of a ball，the earth would then be the size of a grain of sand.

【译文】假定把太阳缩小到小球那么大，那地球就会像沙粒一样小。

1.3.3　祈使句的使用

为了表示指示、建议、劝告和命令等意思，经常采用祈使语气，在使用说明书、操作规程、作业指导、程序建议和注意事项等资料中尤为多见。

例31　If necessary，check if the circuit diagrams and instruction for operation are still applicable.

【译文】必要时，应核对线路图和使用说明书是否仍然适用。

例 32 Be careful not to mix the liquids.

【译文】注意不得把这几种液体混合起来。

有时，为了表示某种设想、假定或条件，也会采用祈使语气。

例 33 Convert heat into mechanical work.

【译文】把热能转换成机械能。

例 34 Open the key, and an induced current in the opposite direction will be obtained.

【译文】如果把电键断开，就会得到反向感应电流。

例 35 Take ten percent off the nonproductive expenses.

【译文】削减非生产性开支 10%。

1.3.4 其他特征

由于科技文章本身的客观性、信息性，在句子结构和其他语言特点上表现出其他一些明显的特点，如分隔结构、非言词符号、倒装、省略的使用等。这些语言特点是应科技英语本身的要求而形成的，既是语法上的要求，而更主要的是出于行文修辞的需要。例如，省略常可节省篇幅，使表述更为简洁和紧凑；倒装常能使某一事理更为醒目和突出，使上下文的联系更为紧密或使描写更为生动；而割裂则往往能使句子的整体结构更为匀称和平衡。

例 36 All bodies consist of molecules and molecules of atoms.

【译文】一切物体都由分子组成，而分子则由原子组成。

原句中第二个 molecules 后省略了 consist。

例 37 All these forces the fuse must be able to withstand without changing its operating characteristics.

【译文】引信必须能承受所有这些力，且其作用性能不得有所改变。

原句中 all these forces 为 withstand 的宾语，形成倒装。

例 38 Thus, it would be correct to say that the distance to the sun, from where we are on the earth, is about 1 million walking days.

【译文】因此，可以这么说，从地球步行到太阳约需一百万天。

原句中 from... earth 与 distance 割裂。

例 39 The most important of the materials in our bodies are the proteins.

【译文】在构成人体的各种物质中，最重要的是蛋白质。

原句中主语 the proteins 与谓语 are the most important 倒装。

例 40 In medical research a relation has been found between measles and such things as behavior problems, personality changes, and dulling of mental ability.

【译文】医学研究已经发现，麻疹与举止反常、性格变态和智力低下等有关。

原句中 between... ability 和 relation 割裂。

总之，科技英语以客观事物为中心，它在用词上讲究准确明晰，论述上讲究逻辑严密，表述上则力求客观，行文上追求简洁通畅，修辞以平实为范，辞格用得很少，句式显得单一少变，语篇中有许多科技词汇和术语以及公式、图表、数字、符号，但句子长而不乱。

翻译练习

1. A body can move uniformly and in a straight line, there being no cause to change that motion.

2. We can store electrical energy in two metal plates separated by an insulating medium. We call such a device a capacitor, or a condenser, and its ability to store electrical energy capacitance. It is measured in farads.

3. Materials to be used for structural purposes are chosen so as to behave elastically in the environmental conditions.

4. Such a slow compression carries the gas through a series of states, each of which is very nearly an equilibrium state and it is called a quasi-static or a "nearly static" process.

5. Computers may be classified as analog and digital.

6. The switching time of the new-type transistor is shortened three times.

7. Notwithstanding clauses (a) and (b) of this Article 5.6 above, the Assigning Party may transfer all or part of its amount of the registered capital of the Company to an Affiliate (the "Affiliate Assignee") of the Assigning Party on the following conditions.

8. Stainless steel, which is very popular for its resistance to rusting, contains large percentage of chromium.

9. Combined with digital television sets, videodiscs can not only present films but also offer surround sound which provides theatre quality-amazing reality by which the viewers may have an illusion that they were at the scene and witnessed everything happening just around them.

10. The center is running a series of talks on the relationship between science and literature, in which poets and scientists discuss how scientific ideas over the past two centuries have influenced literature and social change.

第2章　科技英语的翻译方法、标准和过程

2.1　科技英语的翻译方法

科技英语的翻译方法主要有直译和意译两种。直译强调的是“形似”，主张将原文内容按照原文的形式（包括语序、词序、语气、结构、修辞方法等）用目的语表达出来，同时要求译文语言通顺易懂，表达清楚明白，符合语言规范。如果能够以与原文相同的形式再现与原文相同的内容，就可采用直译的方法。

例1　In some automated plants electronic computers control the entire production line.

【译文】在某些自动化工厂，电子计算机控制整个生产线。

例2　Since the oldest forms of life were all sea life, many scientists believe life began in the sea.

【译文】由于最古老的生命形式是海洋生物，因此许多科学家认为生命起源于海洋。

然而，英语和汉语毕竟是两种不同的语言，有时直译往往行不通。在这种情况下，译者就要考虑怎样摆脱原文的句子结构，用不同的汉语形式来表达原文的意思。因此，必须先吃透原文，在正确理解原意的基础上，重新遣词造句，把原文的意思用通顺的汉语表达出来，这种翻译方法称作“意译”。即打破原文的语言形式，用目的语的习惯表达形式再现原文的意蕴。但形式的转换或再创造必须忠实原文，不能进行编撰或杜撰，不能偏离了原文的内容与风格。

例3　The reason why air makes fire burn more intensely was learned only about two hundred years ago, when several scientists finally proved that oxygen, one of the gases air contains, can combine with certain other elements like carbon to release much heat.

【译文】空气为什么能使火燃烧得更旺？直到大约200年前才弄清其原因。当时某些科学家终于证明，空气中有一种气体叫氧，它能够与其他一些元素（如碳）化合，从而释放出大量的热。

例4　It is easy to compress a gas. It is just a matter of reducing the space between the molecules. Like a liquid a gas has no shape, but unlike a liquid it will expand and fill any container it is put in.

【译文】气体很容易压缩，那只不过是缩小分子之间的距离而已。气体和液体一样没有形状，但又不同于液体，因为气体会扩张并充满任何盛放它的容器。

例5　A large segment of mankind turns to untrammeled nature as a last refuge from encroaching technology.

【译文】许多人都想寻找一处自由自在的地方，作为他们躲避现代技术侵害的世外桃源。

例6　Newton created a planetary dynamics which was so successful that for many years scientists complained that nothing was left to be done.

【译文】牛顿创立的天体力学真是天衣无缝，以致许多年来科学家们抱怨无空可补。

当然，在翻译实践中，完全用直译或完全用意译的情况并不多见。通常，能直译的地方就直译。直译的好处在于既能表达原文的意思，又能尽量保持原文的语言风格。不能直译的地方就采取意译。何时直译何时意译应根据原文的上下文和译文的表达需要而灵活掌握。但是，无论直译还是意译，都要把忠实于原文的内容放在第一位。如果译文只忠实于原文形式而不忠实于原文内容，那就不是直译而是硬译。如果译文只追求通顺的形式而不忠实于原文的内容，那就不是意译而是滥译。

2.2　科技英语的翻译标准

关于翻译的标准，中西方很多翻译家都提出了不同的见解。中国著名翻译家严复提出的“信、达、雅”三标准一直为不少翻译工作者所推崇。具体一点，就是“译文要忠实于原文的思想”、“译文要合乎全民规范化的语言”、“译文要保持原文的风格”。奈达则提出了“Functional Equivalence”（功能对等），要求译文在词汇意义、文体特色等诸层面上尽可能与原文保持一致。

科技英语翻译的标准通常可以用两个字来概括，即“信”与“达”。所谓“信”，指的是译文要忠实于原文，必须符合原意，不得有任何篡改；所谓“达”，指的是译文必须通顺达意，符合语言规范。好的译文，既要忠实于原文的意思和风格，同时读起来又要流畅。与原意大相径庭的文字，不管多么通顺，都算不上是成功的翻译。翻译要在忠实于原文内容的基础上力求译文表达形式上的通顺，又要在译文通顺的前提下尽可能保持原文的形式。如果译文的“通顺”与“忠实原文形式”两者之间出现了矛盾，则不必拘泥于原文的形式。翻译是一种再创作，只有在充分理解原文的基础上灵活运用各种翻译技巧，多种方法相结合，才能提高译文的语言质量，达到既“信”且“达”的境界。

例7　Irrespective of the approach, it is important to agree tariffs before investments proceed.

【译文】不论方法如何，重要的是投资前达成一致同意的电价。

例8　Technical problems were of greatest importance in the case studies where there were weak links between relatively developed grids.

【译文】在两个相对发达的电网之间存在着薄弱的连接时，技术问题在案例研究中最具重要性。

例9　The research work is being done by a small group of dedicated and imaginative scientists who specialize in extracting from various sea animals substances that may improve the health of the human race.

【译文】一小部分富有想象力和敬业精神的科学家正在进行这项研究，他们专门研究从各种海洋动物中提取能增进人类健康的物质。

2.3 科技英语的翻译过程

科技英语的翻译过程可分为理解、表达和核实三个阶段。理解是基础，表达是关键，核实是保证，三者缺一不可。

理解阶段，主要是领会原文内容。译者要弄清每个词语、每个词组、每个单句的确切含义，也要弄清每一句话的结构、逻辑、重点、与上下文的关系、所采用的语气等，即弄清原作的全部精神实质。译者要想把握住原文的全部意图和精神实质，就必须具有对整个文章的布局谋篇进行宏观分析的能力。任何一篇文章或一段文字都是一个有机的整体，词与词、词与句子、句子与段落甚至整个篇章之间，都有着必然的内在联系。在反复通读原文，领略大意之后，译者应透过各种语言现象对句子进行微观分析。在对原文的理解过程中，单纯局限于表层语言现象的理解往往不能解决所有问题，而必须透过原作的文字，准确弄清作者所阐明的事理和事物间的逻辑关系。只有对原文深入体会，彻底理解作者的真实意图，译文所表达的内容和所传递的信息才能跟原作一致，真正达到和作者的交流和沟通，才能根据原文的字面意思做出有把握的推理，并敢于进行恰如其分的发挥，最终获得正确的判断。尤其在科技英语翻译中，理解不仅局限于原文的语言，还要掌握与文章相关的技术背景知识。

在正确理解的基础上，译者需要用清晰明了的语言把作者的意思表达出来，尤其要注意汉语与英语在语言结构方面的差异。在翻译的表达阶段，要特别注意的是，翻译的表达和创作的表达是不一样的。译者所表达的是原文作者已经表达出来的东西，因此必须按照原作者的思维逻辑来表达。虽然允许并提倡在深刻理解原文的基础上创造性地表达，但译者不能任意发挥，随意删改。在此过程中，应该注意表达的规范性和逻辑性。科技文体讲究论证的逻辑性，要求语言规范，概念明确，逻辑严密，表述无懈可击。作为译者，也必须用符合逻辑的语言传达出作者的意图和思想。译者不仅要考虑句中的各种语法关系，更要注意各概念之间的逻辑关系，要从逻辑角度来判断自己的译文是否能准确传达原文所要传递的信息。

第三阶段是核实译文是否准确，表达是否流畅，是否符合汉语的习惯表达，能否被读者接受。要在前两阶段的基础上，对汉语表达字斟句酌，不厌其烦地反复修改和核对。译文的质量没有止境，只有反复修改才会发现自己存在的问题，从而提高翻译能力，提高翻译质量。

例 10　The jobs that the computer has created have helped our economy and standard of living rather than undermined them.

【译文】计算机创造的职业对我们的经济和生活水平起着促进作用而不是破坏作用。

例 11　In any event the reliability factor for cast iron should be evaluated in a laboratory testing program because the variance of the mechanical properties may be quite large.

【译文】由于铸铁机械性能的变化可能相当大，铸铁的稳定系数应在实验室中按实验程序予以确定。

例 12　As an oil ship is loaded, it sinks deeper into the water displacing an additional amount of water equal to the weight of the added load.

【译文】当油船装了货，吃水就深些，排开的水量即等于所加负荷的重量。

例 13　Heat transfer helps to shape the world in which we live. Great loss is suffered by man when heat transfer is impossible. If a way could be found to transfer heat to the polar regions, they could support large populations just as the temperate countries do.

Luckily, man has been more successful in making use of the natural methods of heat transfer on a smaller scale. A quantity of heat is useless if it is where it is not needed. It may be useful if it can be moved to another place. Heat, by its very nature, helps to make this possible. Heat always travels, or flows, from a high temperature to a low temperature.

Conduction, for example, is a point-by-point process of heat transfer. If one part of a body is heated by direct contact with a source of heat, the neighboring parts become heated one after another. Thus if a metal rod is placed in a stove, heat travels along the rod by conduction. The molecules of the metal in the rod increase their energy of motion. This violent motion is passed along the rod from molecule to molecule.

When heat is transferred by the movement of the material which is heated, it is called convection. It applies to free-moving substances. That is, liquids and gases. The motion is a result of changes of density that accompany the heating process. Water in a tea kettle is heated by convection.

【译文】热量传输有助于改变我们所生活的这个世界。如果热量无法传输，那么人类将遭受巨大的损失。假如能够有办法把热量传输到两极地区，那么这些地区就会像温带国家一样，可以有大量人口居住。

幸运的是，人类能够成功地利用自然方法进行小规模的热量传输。若某处存在一部分不需要的热量，那么这部分热量是毫无用处的，然而如果把它移到其他地方或许就能有用。热量本身的特点使这种想法成为可能。因为热量总是在运动中，即从温度高的地方“流动”到温度低的地方。

例如，传导就是热量从一点到另一点的传输过程。如果用热源以直接接触的方式加热一个物体的某一部分，那么其邻近部分也会逐渐热起来。因此如果将一个金属棒放在火炉上，那么热量就会沿着金属棒传导。金属棒中的分子增加动能，产生激烈运动。一些分子的激烈运动传给其他分子，于是热量沿着金属棒传播。

当热量的传输是通过所加热的物体运动来完成时，称为对流方式。这种方式适于能够自由运动的物质，如液体、气体等。随着物体的受热，其密度不断变化，于是形成对流。水壶中的水就是以这种方式加热的。

翻 译 练 习

1. As the mass of a body is constant, the effect of the increase in applied force is to increase the velocity.

2. The best way to improve urban air may be to curb the use of cars, even though modem cars are far cleaner than earlier ones.

3. The free electrons usually do not move in a regular way.

4. Far from increasing the reaction rate, high temperatures decrease it.

5. These pumps are featured by their simple operation, easy maintenance and durable service.

6. Only when a rocket attains a speed of 18,000 odd miles per hour, can it put a manmade satellite in orbit.

7. From just the entrance, in the southern part of the bay, up to a second slack water area, about midway up the bay, tidal currents are ebbing.

8. But now it is realized that supplies of some of them are limited, and it is even possible to give a reasonable estimate of their "expectation of life", the time it will take to exhaust all known sources and reserves of these materials.

9. The dataset consists of twice daily 500-hPa height analyses from the U. S. National Meteorological Center for the 20 winters 1962/1963 through 1981/1982.

10. The chain reaction must be carefully controlled. This is done by absorbing excess neutrons in boron control rods. The heat is transferred by circulating carbon dioxide gas. The hot gas is used to produce steam. The whole assembly is called a nuclear reactor.

第3章 翻译技巧

科技英语大量使用专业词汇和半专业词汇。专业词汇指仅用于某一学科或专业的词汇或术语。每门学科和专业都有自己的一套含义精确而狭窄的术语。从词源角度分析，专业词汇有两个主要来源：一是来自英语日常词汇；二是来自拉丁语和希腊语词根及词缀的词汇。半专业词汇指既用于日常英语，同时又常用于科技英语中的词汇。半专业词汇与专业词汇的主要区别在于半专业词汇一般不专用于某一学科，而是为各学科所通用，这些词汇用在不同学科中虽然基本含义不变，但其确切含义则存在较大差别。比如，power 一词在日常英语中表示“力量、权力”等意思；用于体育专业表示“爆发力”；用于机械专业表示“动力”；用于电力专业表示“电力”；用于物理专业表示“功率”等。

3.1 词义的选择

汉语、英语两种语言都存在一词多义的现象。由于用法习惯不同，在某一特定场合下可以译为另一种语言的某个词，但在其他场合下却不能这么译。翻译的主要任务之一便是确定一个词在一句话中的确切词义，而翻译的失误往往是词的释义挑选、使用不当，或找对了释义而又没有使其前后合理搭配。比如，China is a very big country with a very big population.（中国幅员辽阔，人口众多）。这里有两个 very big，都表示“很大”的意思，用在 country 之前，译为“辽阔”，用在 population 之前，译为“众多”，可见词汇释义与其后的中心词有关，词的搭配不同，词汇释义就不同。在科技英语翻译中，必须注意不同专业中常用的词义，根据具体情况在词典释义基础上进行恰当的“演绎”，选出最确切的词义。

把握词汇的本质词义是选择词义的基础。在众多的词义中，选择出一个最确切的词义是正确理解原文所表达的思想的基本环节，是翻译成功的基础。多义词词义的辨析、选择和确定，是一项艰苦的思维过程，选择时一般可以从以下几个方面着手：

3.1.1 根据词类确定词义

词义的选择就是用最恰当的词汇表达再现原文，这是翻译中最难掌握但又十分重要的部分。英语词汇的意义非常广泛，相对应的汉语意义也十分繁杂，同一个词汇有许多意思、多种解释。要选择正确的词义，首先要确定该词在句中属于哪一种词类，然后再确定词义。在下面各句中，like 一词分属于几个不同的词类，试比较：

例1 Like charges repel; unlike charges attract.

【译文】同种电荷相互排斥，异种电荷相互吸引。

like 是形容词，表示“相同的”。

例2 Like knows like.

【译文】英雄所见略同。

like 是名词，表示“同类人”。

例 3 It is the atoms that make up iron，water，oxygen and the like.

【译文】正是原子构成了铁、水、氧等类物质。

like 是名词，表示“相同之物”。

例 4 We like to study computer science more than chemistry.

【译文】我们喜欢学习计算机科学更甚于化学。

like 是动词，译作“喜欢”。

例 5 Under a microscope，a cell looks like a bit of clear jelly with a thin wall round it.

【译文】在显微镜下，细胞看上去就像一丁点清澈的覆有薄膜的胶状物。

like 是介词，译作“像”。

例 6 The actual interest rate is more like 18 percent.

【译文】实际利率更接近 18%。

like 是副词，译作“接近”。

例 7 It sounds to me like they don't know what water is made of.

【译文】听他们的口气似乎他们不知道水是由什么构成的。

like 是连词，译作“似乎”。

再举两个例子：

例 8 Capacitors are devices，the principal characteristic of which is capacitance.

【译文】电容器是以电容为主要参数的元件。

例 9 The narrow，directive beam is characteristic of radar antennas.

【译文】窄定向波束是雷达天线所特有的。

例 8 中，characteristic 为名词，译为“参数”；例 9 中，characteristic 为形容词，译为“特有的”。

3.1.2 根据学科分支和专业选择词义

在英语和汉语中，同一个词即使属于同一词类，在不同的学科领域或不同的专业中往往具有不同的词义。在科技翻译中，首先要明确原文所涉及的学科范畴，然后再根据专业内容确定词义，这样不仅可以迅速地缩小词义的选择范围，而且可以提高所选词义的精确度。比如，introduction 在一般用语中的词义是“介绍，就职”，但在物理学范畴词义就变成了“感应线圈”。在选择词义时应尽量从专业实际出发，采用国家技术标准中规定的术语及权威性的专业词典，避免使用行业俗语和非标准性专业术语。翻译时必须充分考虑词汇的使用场合来确定词义。

例 10 The lathe should be set on a firm base.

【译文】车床应安装在坚实的底座上。

例 11 A base of logarithms must be a positive number，not equal to one.

【译文】对数的底数必须是一个不为一的正数。

例 12 When an acid reacts with a base，a neutral product usually results，and we call it a salt.

【译文】当一种酸和一种碱起反应时，通常形成一种中性产物，我们称之为盐类。

例 13　Line AB is the base of the triangle ABC.

【译文】AB 线是三角形 ABC 的底边。

例 14　A transistor has three electrodes, the emitter, the base and the collector.

【译文】晶体管有三个电极，即发射极，基极和集电极。

base 一词的基本含义是“基础”，是一个概括性较强的词。但在以上五句中，都因不同的学科分类而转化为不同的含义。在例 10 中，属于机械方面的术语，译为“底座”；在例 11 中为数学术语，译作“底数”；例 12 中属于化学术语，译为“碱”；例 13 则为几何术语，译为“底边”；在例 14 中，属于电子术语，译为“基极”。

例 15　Currently the press, paint and assembly shops operate two shifts, while forging is operating three.

【译文】目前，冲压、涂装、装配车间实行两班制，锻造车间实行三班制。

press 一词的基本含义为“印刷机，新闻，压”等，是一个有不同词性的词。在译文中，因应用场合和技术内容不同而译成“冲压”。而 paint 在这里不是“绘画”，而译为“涂装或喷漆”。

例 16　Schematic diagram of NiCad battery charger suitable for mobile use. See text for explanation of DS1 and D3. D2 protects the components in the event of accidental reversal of input leads.

【译文】这是镍镉电池充电器电路图，这个充电器适合移动使用。DS1、D3 的参数见文。D2 能够防止万一引线接反损坏元件。

译文中将 leads 译为“引线”，而不应根据文中 battery charger suitable for mobile use，就望文生义，认为 leads 是指“铅”或“铅制汽油”。

例 17　Relaying of single bus is relatively simple since the only requirements are relays on each of the circuit plus a single bus relay.

【译文】因仅需对每条线路以及母线进行继电保护，所以单母线的继电保护相对较简单。

在这里 bus 用在了与科技英语相关的文献中，所以应选择与其相关的术语，因此译为“母线”。

例 18　In the electrolysis, we must make use of cell.

【译文】在电解过程中，我们必须使用电解槽。

例 19　When the ends of a copper wire are joined to a device called an electric cell, a steady stream of electricity flows through the wire.

【译文】当把一根铜丝的两端连接到一种叫做“电池”的电器上时，就会有稳定的电流流过该铜丝。

例 20　The nucleus is the information center of the cell.

【译文】细胞核是细胞的信息中枢。

例 18～20 中的 cell 分别用于化学、电学和生物学领域，根据具体情况分别译为“电解槽”、“电池”和“细胞”。

3.1.3 根据搭配习惯选择词义

英语的一词多义也体现在词与词的搭配上。一般说来，一个孤立的词其词义是不确定的，但当处于特定的搭配关系时，词义就变得明朗化了，不同的搭配方式，可以产生不同的词义。以 resistance 为例：

oil resistance 耐油性
resistance of materials 材料力学
resistance of/to deformation 变形抗力
resistance to heat 耐热性
resistance to impact 抗冲击强度
resistance to oxidation 抗氧化力
resistance to pressure 耐压性
resistance to sparking 击穿电阻
resistance to traction 牵引阻力
resistance to wear 耐磨性
weather resistance 耐风雨侵蚀能力

例 21 For the years there's been a labor force deficit in the industry.

【译文】许多年来，这个行业一直存在着劳动力短缺问题。

industry 的含义不止一个，如，advertising industry 广告行业，service industry 服务业，automobile industry 汽车业，tourist industry 旅游业，education industry 教育产业，等等。若将原句中的 industry 译为“工业”，则与“劳动力短缺”搭配不当，因为“工业”自身包含了许多种，不大可能一概缺乏劳动力，因此，译文中将“工业”演绎为“行业”。

例 22 The submarine cable here turns to the right.

【译文】这儿的海底电缆是向右转弯的。

例 23 The factory turns out 100 automobiles in a day.

【译文】这个工厂每日生产汽车 100 辆。

以上两个例句中的动词是 turn，但一个搭配介词 to，译为“拐弯”，另一个搭配副词 out，译为“生产”。

例 24 The brightness of a spot got by the middle of the tube can be varied by adjusting voltage between grid and cathode.

【译文】显像管中央得到的光点亮度可以通过调整栅极和阴极之间的电压加以改变。

3.1.4 根据上下文和语境选择词义

任何语言现象都不是孤立存在的。英语中有两句话：“No context，no text”，“You know a word by the company it keeps”。这里的 context 和 company 既指“上下文”和“搭配”，又指“具体情境”或“语境”。翻译家庄绎传先生曾经说：“一个词用在不同的场合会有不同的含义，译者不能只想到自己最熟悉的那个含义，而要充分利用上下文，依靠能够获得的相关信息，判断出词的确切含义。”由于英语具有一词多义特点，每个词语的上下文多有变化，并且由于其固定或半固定的不同搭配，在译文中往往需要使用不同的词语与之对应，而且应当非常具体。如，Tension is building up. 根据不同的上下文，可视情况分别译作：形势紧张起来；张力在增大；电压在增加；压力在增加；血压在增高。比较好的做法是从整个句子甚至段落的层面上来判断，借助于上下文提供的各种线索做出合理的分析、推理、判断，以选择适当的词进行表达。以 develop 一词为例：

例 25　Albert Einstein, who developed the theory of relativity, arrived at this through mathematics.

【译文】创立相对论的阿尔伯特·爱因斯坦是通过数学得出这一理论的。

例 26　Shorts frequently develop when insulation is worn.

【译文】绝缘外壳磨破后往往会发生短路。

例 27　After the war much of this knowledge was poured into the developing of the computers.

【译文】战后，这项知识大量地应用于研制计算机。

例 28　Most of the money came from selling the secret of a new type of potato he had developed.

【译文】这笔钱大都来自他出售培育新品种马铃薯的秘方。

例 29　Packaged software is developed to serve the specific needs of one user.

【译文】软件包的开发只是为某一用户的特定需要服务。

例 30　In developing the design, we must consider the feasibility of processing.

【译文】在进行设计时，必须考虑加工的可能性。

例 31　Other isolation methods are being developed.

【译文】目前正在研究其他隔离方法。

例 32　Noises may develop in a worn engine.

【译文】磨损的发动机里可能会产生各种噪音。

例 33　Sure enough, 90 percent of the plants developed the disease.

【译文】可以肯定，90% 的庄稼都染上了这种病。

例 34　Developing the potentialities of atomic energy, are all examples of the work of the applied scientist or technologist.

【译文】挖掘原子能的潜力等，这都是应用科学家和技术人员的工作。

例 35　I developed a photograph in the nearby shop.

【译文】我在隔壁的商店冲洗照片。

develop 是一个词义丰富，使用频率很高的单词，它与不同的名词搭配会有不同的译法。在以上的例句中，根据不同的语境，动词 develop 在不同的句子中分别译成了“冲洗”、“发生”、“培育”、“开发”和“创立”等。

再以 condition 一词为例：

例 36　A mathematical or logical operation must meet a certain condition.

【译文】数学和逻辑运算必须满足某种条件。

例 37　Interactive programming conditions are now available for some commercial programming languages.

【译文】交互编程环境现在能用于一些商用编程语言。

例 38　The results of a biopsy indicate a rare nonmalignant condition.

【译文】活组织检查结果表明这是一个罕见的良性病例。

例 39　The frequency with which the filter should be removed, inspected, and cleaned will be determined primarily by aircraft operating conditions.

【译文】过滤器拆卸、检查及清洗的次数主要取决于飞机的运行状况。

例 40 The goods have been unloaded into the carrier's house. As they are in such a damage condition, we doubt we will be able to take delivery.

【译文】货物已经卸到公司仓库。由于它们破损如此严重，我们恐怕难以提货。

例 41 There are time and personal health condition to be considered. For example, green tea is good in summer. It seems to dispel the heat and bring on a feeling of relaxation. However, it is not proper for pregnant women to drink green tea.

【译文】要考虑时节和个人身体状况，比如，绿茶适宜在夏季喝，它似乎能够驱散炎热并带来放松的感觉。然而，孕妇不适合喝绿茶。

例 42 The merger offer I think is inadequate considering the current financial condition of this company.

【译文】考虑到这家公司目前的财政状况，我认为该合并提议是不充分的。

例 43 Nonalcoholic fatty liver disease. Medically, the liver transplant was impossible with her condition.

【译文】非酒精性脂肪肝。医学上说，她当时的条件是不可能进行肝脏移植的。

3.2 词类的转换

由于英汉表达方式和习惯不同，句法结构也不同，英译汉时，常常遇到一些无法直译的词或词组，这时应该根据上下文和原文的字面意思，做适当的词类转换。因此，词类转换是英译汉中最常见的方法。

3.2.1 动词的转换

3.2.1.1 动词转换为名词

例 44 The inflammation is characterized by red, swelling, fever, and pain.

【译文】炎症的特点是红、肿、热、痛。

原文中 characterize 为动词，汉语译文中将其译成了名词“特点”。

例 45 A material balance is based on the law of conservation of matter.

【译文】物质平衡是以物质守恒定律为基础的。

例 46 Coating thickness ranges from 0.1 mm to 2 mm.

【译文】涂层厚度范围为 0.1 到 2 毫米。

例 47 The new medicine will expire in 2 years.

【译文】新药有效期为两年。

例 48 Gases differ from solids in that the former have greater compressibility than the latter.

【译文】气体和固体的区别在于前者比后者有更大的压缩性。

3.2.1.2 动词转换为副词

当英语句子中的谓语动词后面的不定式短语或分词转换成汉语句子中的谓语动词时，原来的谓语动词就相应地转换为汉语的副词。

例 49 The molecules continue to stay close together, but do not continue to retain a regular fixed arrangement.

【译文】分子仍然紧密地聚集在一起，但不再继续保持有规律的固定排列形式。

例50 These products tend to react with soap and detergents and produce soap scum.

【译文】这些产品往往会跟肥皂和洗涤剂发生反应，进而产生肥皂浮渣。

3.2.2 名词的转换

英语中名词的使用非常频繁，大量使用名词是英语的一个明显特点。在很多情况下，汉语中采用动词、形容词和副词等形式表述的某个想法或概念，在英语中却通常用名词或动名词来表示。所以当把英文翻译成中文时，往往将原文中的名词或动名词转换成动词、副词和形容词等。

3.2.2.1 名词或动名词转换为动词

汉语中表示行为和动作时多用动词，而名词是英语中最常用的表达方法。因此，英译汉时把英语名词转译成汉语动词的现象较多。

（1）名词转换为动词

英语中有大量的由动词派生的名词和具有动作意义的名词，在翻译过程中这类名词通常转译成动词。

例51 With this software so available there seems little need for analysts to develop their own programs.

【译文】这个软件非常适用，看来程序分析师不需要自己再编新程序。

例52 To design is to formulate a plan for the satisfaction of a human need.

【译文】设计就是议定某种方案以满足人们某种需求。

例53 China's successful explosion of its first atom bomb caused tremendous repercussion throughout the world.

【译文】中国成功地爆炸了第一颗原子弹，在全世界引起了巨大的反响。

例54 The construction of a scientific theory may be compared to the preparation of a weather map at a central meteorological station.

【译文】创立科学理论的工作可以同在中心气象台制作气象图相比。

例55 In a similar way, in the theory of light, we use terms like "waves" and "particles" for the description and discussion of the results of experiments.

【译文】与此类似，在光学理论中，我们采用诸如“波”和“粒子”之类的术语来描述和讨论实验结果。

（2）动名词转换为动词

例56 One of our ways for getting heat is by burning fuels.

【译文】我们获得热的一种方法是燃烧燃料。

例57 An understanding of the laws of friction is important in the designing of modern machines.

【译文】了解摩擦的规律对于现代机器的设计是重要的。

例58 Heating water does not change its chemical composition.

【译文】把水加热不会改变水的化学成分。

例 59 The flowing of current first in one direction and then in another makes an alternating current.

【译文】电流先向一个方向流动，然后又向另一方向流动构成交流电。

3.2.2.2 名词转换为形容词

英语中有些用作表语的抽象名词，其修饰词为不定冠词，翻译成汉语时常常会译为形容词。

例 60 This experiment is an absolute necessity in determining the best processing route.

【译文】对确定最佳工艺流程而言，这次实验是绝对必要的。

例 61 I am a stranger to the operation of electronic computer.

【译文】我对电子计算机的操作是陌生的。

例 62 His experiment was a success.

【译文】他的试验是成功的。

例 63 I found a lot of difficulties to continue the experiment.

【译文】我觉得要继续进行试验是非常困难的。

3.2.3 形容词的转换

3.2.3.1 形容词转换为动词

英语中有些表示知觉、情绪、欲望等心理状态的形容词作表语时常可转换成汉语动词。这类形容词有：able、acceptable、afraid、angry、anxious、ashamed、aware、careful、cautious、certain、concerned、confidant、content、delighted、doubtful、familiar、glad、grateful、ignorant、sorry、sure、thankful、uncertain 等。

例 64 If we were ignorant of the structure of the atom, it would be impossible for us to study nuclear physics.

【译文】我们如果不知道原子的结构，就不可能研究核子物理学。

例 65 Once inside the body, the vaccine separates from the gold particles and becomes active.

【译文】一旦进入体内，疫苗立即与微型金属粒子分离并激活。

例 66 Both of the substances are not soluble in water.

【译文】这两种物质都不溶于水。

例 67 Are you familiar with the performance of this type of transistor amplifier?

【译文】你熟悉这种晶体管放大器的性能吗?

3.2.3.2 形容词转换为名词

英语形容词转换为名词有两种情况：英语中有些形容词加上定冠词表示某一类的人或物，一般作复数名词，翻译成汉语时常译成名词；英语某些表示事物特征的形容词作表语时可转译成名词，其后往往加上“性”、“度”、“体”等。

例 68 Whenever one body touches another, heat always passes by conduction from the warmer to the colder.

【译文】当一个物体接触到另一个物体时，热量总是从较热的物体传导到较冷的物体。

例 69 This workpiece is not more elastic than that one.

【译文】这两个工件都没有弹性。

例 70　In fission processes the fission fragments are very radioactive.

【译文】在裂变过程中，裂变碎片的放射性很强。

例 71　Glass is much more soluble than quartz.

【译文】玻璃的可溶性比石英大得多。

例 72　This metal is less hard than that one.

【译文】这种金属的硬度比那种差。

3.2.3.3　形容词转换为副词

形容词转换成副词最为常见的有以下两种情况：

（1）当英文中的名词转换为汉语的动词后，原来修饰名词的形容词或分词，相应地转换为副词。

例 73　A very brief description of the behavior of materials is provided in the essay.

【译文】在本文中我们简要地阐述了这些材料的性能。

例 74　We must make full use of existing scientific technology.

【译文】我们必须充分利用现有科学技术。

例 75　All of this proves that we must have a profound study of properties of proteins.

【译文】所有这一切证明，我们必须深入地研究蛋白质的特性。

例 76　Successful mass production depends on close inspection and strict control of the quality of manufactured products.

【译文】成功地进行大批量生产取决于对产品质量的严格检测和控制。

例 77　Below 4 °C, water is in continuous expansion instead of continuous contraction.

【译文】在 4 °C 以下，水不是不断地收缩，而是不断地膨胀。

（2）在英语中作表语的名词转换成汉语的形容词，原文中修饰名词的形容词就相应地转换成汉语的副词。

例 78　This experiment is an absolute necessity in determining the solubility.

【译文】对确定溶解度来说，这次试验是绝对必要的。

原句中的名词 necessity 转换成汉语中的形容词“必要的”，形容词 absolute 转换成汉语中的副词“绝对”。

例 79　Gene mutation is of great importance in breeding new varieties.

【译文】在新品种培育方面，基因突变是非常重要的。

3.2.4　副词的转换

3.2.4.1　副词转换为形容词

在翻译过程中，由于原文中副词所修饰的词在翻译成汉语时发生了词类的转换，所以原文中的副词在汉语译文中也应进行相应的转换。主要有以下几种情况：

（1）当原文中的副词所修饰的动词转换为汉语里的名词时，该副词在译文中相应转换为形容词。

例 80　The capacity to move information quickly and inexpensively is due to entirely new uses of communications.

【译文】快速廉价的信息传递是由于通讯设备彻底更新的原因。

原文中副词所修饰的动词 move 转换成了汉语里的名词“传递”，所以用来修饰动词的副词 quickly 和 inexpensively 就相应地转换成了形容词。

例 81 The wide application of electronic computers affects tremendously the development of science and technology.

【译文】电子计算机的广泛应用，对科学技术的发展有极大的影响。

例 82 Earthquakes are closely related to faulting.

【译文】地震与断层的产生有密切的关系。

例 83 The electronic computer is chiefly characterized by accuracy and quick computation.

【译文】电子计算机的主要特点是运算正确迅速。

（2）当原文中的形容词转换为汉语的名词时，原来修饰形容词的副词往往转换成形容词。

例 84 Gasoline is appreciably volatile.

【译文】汽油具有很强的挥发性。

例 85 It is demonstrated that dust is extremely hazardous.

【译文】已经证实，粉尘具有极大的危害。

（3）英语中副词用在名词前或后，在意义上相当于定语，译成汉语时这种副词变为形容词。

例 86 The power plant supplies the inhabitants sixty *li* about with electricity.

【译文】这个电厂供电给周围六十里的居民。

例 87 The buildings around are mostly of modem construction.

【译文】附近的建筑物大部分是现代化的。

3.2.4.2 副词转换为名词

例 88 The attractive force between the molecules is negligibly small.

【译文】分子间的吸引力非常小，可以忽略不计。

例 89 All structural materials behave plastically above their elastic range.

【译文】超过弹性极限时，一切结构材料都会显示出可塑性。

例 90 These parts must be proportionally correct.

【译文】这些零件的比例必须准确无误。

例 91 Chlorine is very active chemically.

【译文】氯的化学性能很活跃。

3.2.4.3 副词转换为动词

英语中有些副词用作表语、状语或宾语补足语，在翻译时往往译作汉语中的动词。

例 92 Open the valve to let air in.

【译文】打开阀门，让空气流进。

例 93 Their experiment has been over.

【译文】他们的试验已经结束。

例 94 In this case the temperature in the furnace is up.

【译文】在这种情况下，炉温就升高。

例 95　If one generator is out of order, the other will produce electricity instead.

【译文】如果一台发电机发生故障，另一台便代替它发电。

3.2.5　介词的转换

3.2.5.1　介词或介词短语转换为动词

有些具有动词意味的介词，在翻译成汉语时需要译成动词。这类介词有：across、along、around、by、for、in、into、over、past、through、throughout、toward、with 等。

例 96　The government is behind this project.

【译文】政府支持这个项目。

例 97　This computer is of high sensibility.

【译文】这台计算机具有很高的灵敏度。

例 98　Without steel, there would be no modern industry.

【译文】如果没有钢，就不会有现代化工业。

例 99　This is against the regulations for operations.

【译文】这是违反操作规程的。

例 100　We are all in favor of this program.

【译文】我们全体赞成这个计划。

例 101　Possibly the most significant research now being conducted is in the use of the laser beam in telephone communications.

【译文】也许现在进行的最重要的研究工作就是利用激光进行电话通信。

3.2.5.2　介词短语转换为名词

例 102　A lot of resources in our country have not been exploited.

【译文】我国有许多资源尚未开发。

例 103　The new precision machine tools are very good in quality and fine in shape.

【译文】新型精密机床的质量很好，样式美观。

例 104　This shows that air is necessary for burning.

【译文】这就证明，燃烧需要空气。

3.3　增　词

英汉两种语言由于表达方式不同，在翻译过程中，往往出现词的添加和减少的现象。所谓增词，就是要根据上下文的意思、逻辑关系以及表达习惯，灵活增加词量，以表达原文字面没有出现但实际已经包含的意思。增词的目的，或是为了补足语气，或是为了连接上下文，有时也是为了避免译文意义含混。增词的情况主要有以下几种：

3.3.1　增加名词

3.3.1.1　增加概括性的名词

在翻译过程中有时需要增加概括性的词语，才能使汉语的语义更加明确。

例 105　The bond between mathematics and the life sciences has been strengthened by the emergence of a whole group of applied mathematics specialties, such as biometrics, psychometrics, and economics.

【译文】生物统计学、心理测验学、计量经济学等一大批应用数学专业学科的出现，大大加强了数学和生命科学之间的联系。

原句中并没有概括词，而翻译成汉语时加上了概括词“等”。

例 106 The frequency, wave length, and speed of sound are closely related.

【译文】频率、波长和声速三方面是密切相关的。

例 107 The advantages of the recently developed composite materials are energy saving, performance efficient, corrosion resistant, long service time, and without environmental pollution.

【译文】最新开发的复合材料具有节能、性能好、抗腐蚀、寿命长和无污染等五大优点。

例 108 Radiant, electrical and chemical energies can all be turned into heat.

【译文】辐射能、电能和化学能这三种能均可转变为热能。

3.3.1.2 在抽象名词后增加名词

英语中有些具有动作意义的或由动词和形容词派生出来的抽象名词，翻译成汉语时一般在其后面增添适当的名词，以使译文意思明确，更符合汉语的表达习惯。

例 109 The temperature needed for this processing is lower than that needed to melt the metal.

【译文】这种加工方法所需的温度要低于熔化该金属的温度。

例 110 Oxidation will make metals rusty.

【译文】氧化作用会使金属生锈。

例 111 In rapid oxidation a flame is produced.

【译文】在快速氧化过程中会产生火焰

例 112 These principles will be illustrated by the following transition.

【译文】这些原理将由如下演变过程说明。

英语中有许多类似的抽象名词。下列抽象名词在翻译成汉语时，要增加括号中的词语，才能使语句更加通顺：activation 活化（作用）、backwardness 落后（状态）、complacency 自满（情绪）、controls 控制（装备）、darkness 漆黑（一团）、development 开发（工程）、distribution 分布（状态）、hysteresis 滞后（现象）、interpolation 内插（法）、ionization 电离（作用）、madness 疯狂（行为）、measurements 测量（结果）、modification 修改（方案）、neutralization 中和（作用）、observation 观察（结果）、preparation 准备（工作）、processing 加工（方法）、resolution 解决（办法）、solution 解决（办法）、termination 终端（设备）、transition 演变（过程）。

3.3.1.3 增加省略的名词

英语中省略名词的部分，在翻译成汉语时要增加必要的名词，语句才能通顺。

例 113 The best conductor has the least resistance and the poorest has the greatest.

【译文】最好的导体电阻最小，最差的导体电阻最大。

例 114 Forces can be classified as internal and external.

【译文】力可以分为内力和外力两种。

例115 Economic globalization has widened the gap between the North and the South and between the rich and the poor.

【译文】经济全球化使南北的发展差距、贫富差距进一步扩大。

3.3.2 增加动词

在英文中有些动词被省略掉了，翻译成汉语时通常需要补充动词。

例116 Matter can be changed into energy, and energy into matter.

【译文】物质可以转化为能，能也可以转化为物质。

例117 The generation plant, transmission lines, and primary substations are shown above the dashed line; the load and distribution below the line.

【译文】发电厂、输电线路、变电站表示在虚线上方，负荷和配电表示在虚线下方。

例118 Science demands men of great effort and complete devotion.

【译文】人们要掌握科学，必须做出巨大的努力并对之怀有无限的热爱。

3.3.3 增加数量词

在英语科技文章的翻译中常常会遇到一些数字的翻译。英语和汉语的表达习惯不同，英语中没有量词，所以翻译成汉语时需要增加量词。比如，first thing　第一件事；the first oil well　第一口油井；a horse　一匹马；a snake　一条蛇；a tree　一棵树等。

例119 The new products will soon be put into use.

【译文】这批新产品即将投入使用。

例120 There are three larger injection machines in the workshop.

【译文】这个车间有三台大型注塑机。

有时还需要增加表示复数概念的数量词，如"许多"、"若干"、"一些"、"一批"、"有些"、"各"、"似"、"诸"等。

例121 Note that the words "velocity" and "speed" require explanation.

【译文】请注意，"速度"和"速率"这两个词需要解释。

例122 A data processor can issue address and function codes.

【译文】数据处理器能发出各种地址码和功能码。

例123 Air is a mixture of gases.

【译文】空气是多种气体的混合物。

例124 For reasons the alternating current is more widely used than the direct current.

【译文】由于种种原因，交流电比直流电用得更为广泛。

3.3.4 增加关联词

英语中的并列连词和从属连词大都以单个词语出现，而汉语中的连词却都是成对出现的。因此，在将英语译成汉语时，应将汉语的连词补充完整，这样才能使译文符合思维逻辑，顺畅而自然。比如，if..., ...（如果……，那么……）；because..., ...（因为……，所以……）；although..., ...（虽然……，但是……）；unless...（除非……，否则……）等。

例 125 Since air has weight, it exerts force on any object immersed in it.

【译文】因为空气具有重量，所以处在空气中的任何一个物体均会受到空气的作用力。

例 126 However carefully boiler casings and steam pipes are sealed, some heat escapes and is lost.

【译文】虽然锅炉壳与蒸汽管是严密封闭的，但是还会有一部分热损耗掉。

例 127 If X is equal to Y, X plus A equals Y plus A.

【译文】若 $X = Y$，则 $X + A = Y + A$。

有时英文句子中并没有出现关联词，但是翻译成汉语时要增添必要的关联词。

例 128 Being large or small all magnets behave the same.

【译文】所有磁铁，无论大小，其性质都一样。

例 129 Carbon combines with oxygen to form carbon oxides.

【译文】碳与氧化合后便构成各种碳氧化合物。

3.3.5 增加修饰性的词

在英译汉的过程中，有时候在译文里增加一些词，能使表达更加确切，或使译文更符合汉语语法和修辞的要求。最常见的有以下几种情况：

3.3.5.1 增加具有修辞作用的名词

例 130 These early cars were slow, clumsy, and inefficient.

【译文】这些早期的汽车速度缓慢，行动笨拙，效率不高。

译文中在形容词前分别加上了三个名词："速度"、"行动"、"效率"，既使译文意思明确，又使其形成四字词组，匀称整齐。

例 131 Science and technology are developing rapidly.

【译文】科学技术日新月异。

例 132 A new kind of computer——small, cheap, fine——is attracting increasing attention.

【译文】一种新型的计算机越来越引起人们的注意——这种计算机体积小巧、价钱低廉、性能优越。

3.3.5.2 增加具有修饰作用的副词

例 133 Inflation has now reached a serious level。

【译文】通货膨胀现在已经发展到空前严重的地步。

译文中增加了"空前"二字，使句子更加鲜明突出，达到了很好的修辞效果。

例 134 It must have been surprising to see a little girl working at a high table, surrounded by maps and all kinds of instruments。

【译文】看到一个小女孩趴在堆满地图和仪器的高桌上聚精会神地工作时，谁都会感到惊讶。

3.3.5.3 其他修辞性增词

例 135 Heat from the sun stirs up the atmosphere, generating winds.

【译文】太阳发出的热能搅动大气，于是产生了风。

例 136 In general, all the metals are good conductors, with silver the best and copper the second.

【译文】一般来说，金属都是良导体，其中以银为最好，铜次之。

例 137　In addition to the speed of erection，these types usually have other advantages.

【译文】除了安装速度快外，这种类型的结构通常还具有其他优点。

3.4　省　略

省略译法通常是将翻译出来反而显得多余累赘的词语省略不译。也就是说，把一些可有可无的，或者有了反显累赘或违背译文表达习惯的词语删去，从而使译文简洁，符合汉语的表达习惯。省略法的目的是保证译文简捷明快、严谨精炼。省译后的译文虽然在词量上和原文不尽一致，但在意思和精神上却应该和原文保持一致。

3.4.1　冠词的省略

英语中冠词的基本功能是语法功能，而非语义功能，冠词本身不具备独立词义。由于汉语中没有冠词，英译汉时，为使译文更具可读性，常常采用省略译法。

3.4.1.1　不定冠词的省略

例 138　He is expected to be an electrical engineer.

【译文】他有望成为电气工程师。

例 139　What is the best time to introduce a new generator set?

【译文】什么时候是引进新发电机组的最佳时机?

例 140　Whether it is a solid，a liquid，or a gas，any substance is made of atoms.

【译文】无论是固体、液体、还是气体，任何物质都是由原子构成的。

这三个例子译文中将不定冠词 an 和 a 省略。

另外，在某些固定词组中的不定冠词汉译时也应省略。

例 141　In a word，power can be transmitted to distant places via cables.

【译文】总之，可以通过电缆传到很远的地方。

类似的固定词组还有：a number of（几个，若干）；a lot of（许多）；a great deal of（许多）；a few（几个）；a little（少许）；as a rule（按规定）；on a large scale（大规模地，大范围地）；as a matter of fact（实际上）；have a mind to（记住），等等。

3.4.1.2　定冠词的省略

例 142　The crocodile belongs to the reptile.

【译文】鳄鱼属于爬行动物。

例 143　The rate of a chemical reaction is proportional to the concentrations of the reacting substances.

【译文】化学反应的速度与反应物的浓度成正比。

例 142、例 143 的译文中均省略了表示类别的定冠词 the。

例 144　The water in Qinghai Lake contains a considerable amount of salt.

【译文】青海湖的水含有大量盐分。

例 145　The inner and outer rings are graded and stored according to size.

【译文】内外圈是根据尺寸分类存放的。

例 144、例 145 的译文中均省略了带限定性定语名词前的定冠词。

例 146 The atom is the smallest particle of an element.

【译文】原子是元素的最小粒子。

例 147 The fourth section of the second chapter concerns the functions of the triode.

【译文】第二章第四节讨论三极管的功能。

例 146、例 147 译文中省略了在形容词最高级和序数词前的定冠词 the.

例 148 The more compressed it is, the hotter the gas becomes.

【译文】越压缩，气体越热。

“the more..., the more...”句式中的定冠词 the 也应省略不译。

另外，某些固定词组中的定冠词也常常省略不译，比如，for the time being（目前）；by the way（顺便说）；on the whole（总之），等等。

3.4.2 代词的省略

3.4.2.1 人称代词的省略

（1）英语中的人称代词 we、you 等及不定代词 one 在句子中作主语时，往往含有泛指的意思，所以在译成汉语时可以省略。

例 149 Since the airplane's mass is not given, we can find it by using this formula.

【译文】既然飞机的质量没有给出，那么可以用这个公式求出来。

例 150 When designing a gear set one cannot consider torque alone.

【译文】设计齿轮组时只考虑扭矩是不够的。

例 151 If you know the wave length, you can find the frequency.

【译文】如果知道波长，就可求出频率。

例 149 ~ 例 151 中的人称代词 we、one 和 you 均译为“任何人，人们”，在这里有泛指意义，可以省略不译。

（2）英语中有些作宾语的人称代词，不管前面是否已经提到过，翻译成汉语时都可以省略。

例 152 A battery has within it some resistance called internal resistance.

【译文】电池内的电阻称为内阻。

例 153 It's your pen. I found it on the playground.

【译文】这是你的钢笔，我在操场上找到的。

例 152 译文中作介词 within 的宾语的人称代词 it 被省略；例 153 译文中作宾语的 it 省略。

3.4.2.2 物主代词的省略

为了清楚地表达从属关系，英语中的物主代词应用频繁，而汉语中却很少使用。所以，英语句子中的物主代词在汉译时往往可以根据汉语的语言习惯省略不译。比如，I washed my hands before having my dinner. 汉语只需说“我吃午餐前洗了手”，而不用说“我吃我的午餐前洗了我的手”。

例 154 The diameter and the length of wire are not the only factors to influence its resistance.

【译文】导线的直径和长度不是影响电阻的惟一因素。

例 155　Different electronic instruments differ in their property.

【译文】不同的电子仪器具有不同的性能。

例 156　Insulators in reality conduct electricity but, nevertheless, their resistance is very high.

【译文】绝缘体实际上也导电，但电阻很高。

例 154 中的 its 根据汉语习惯省略不译；例 155、例 156 的译文中省略了对物主代词 their 的翻译。

3.4.2.3　代词 it 的省略

it 作为代词除了可以指代前面所提到的事物以外，还可能有以下三种特殊用法：① 用作无人称指代，如指代天气、时间等；② 在强调结构中用作主语；③ 用作形式主语或形式宾语。这三种特殊用法中的 it 在汉译时往往省略不译。

例 157　It is ten o'clock sharp now.

【译文】现在是十点整。

例 158　It is cold and windy outside.

【译文】外面又冷风又大。

例 157 和例 158 中分别指代时间和天气的 it 在译文中均省略。

例 159　It was not until the 20th century that man realized the importance of the solar energy.

【译文】直到 20 世纪人类才认识到太阳能的重要性。

例 160　Reading furnishes the mind only with materials of knowledge; it is thinking that makes what we read ours.

【译文】阅读只能用知识材料装备头脑，而思考才会变我们之所读为我们之所有。

例 159 和例 160 中的 it 作为强调结构的主语，汉译时均省略。

例 161　It is assumed that the load is uniformly distributed between the welds.

【译文】假设两焊缝间的载荷为均匀的。

例 162　It is of course true that the working stresses should be below the elastic limit of the material.

【译文】材料的工作应力应低于其弹性极限。

例 163　Experts consider it possible to cure various diseases via radium radiation.

【译文】专家们认为通过镭辐射治疗各种疾病是可能的。

例 164　This formula makes it easy to determine the wave length of sounds.

【译文】这一公式使得求声音的波长十分容易。

例 161 和例 162 中 it 作形式主语，例 163 和例 164 中 it 作形式宾语，汉译时均可省略不译。

3.4.2.4　反身代词的省略

反身代词在英语中用作宾语或同位语的居多，在汉译时可根据汉语的语言习惯省略不译。

例 165 Finally, this condensate is pumped back into the boiler and the cycle repeats itself.
【译文】最后，冷凝水被泵回锅炉，这种循环重复进行。
例 166 We should concern ourselves here only with the stability of the new system.
【译文】在此，我们只讨论新系统的稳定性。
例 165 中的 itself 和例 166 中的 ourselves 在汉译时均可省略。

3.4.3 介词的省略

英语中词与词、词组与词组之间的关系常通过介词表示，而汉语中句子成分之间的关系常靠词序和逻辑关系体现出来。因此，英语中的许多介词常省略不译。由于英语介词丰富，而汉语介词数量有限，英译汉时介词除转译为汉语动词等词类外，省去不译的情况也比较多见。

3.4.3.1 省略表示时间或地点的介词

例 167 The first electronic computer was produced in our country in 1958.
【译文】我国第一台电子计算机是 1958 年生产的。
例 168 This engine has being working for over 10 hours.
【译文】这台发动机已经连续工作超过 10 小时。
例 167 中表示时间和地点的两个介词 in 均被省译；例 168 中表示时间的介词 for 被省译。
例 169 Smoking is not allowed in the store-house.
【译文】仓库重地，不准吸烟。
例 170 Whenever a current flows through a resistance, a potential difference exists at the two ends of the resistance.
【译文】电流通过电阻时，电阻的两端就有电位差。
例 169 译文中引导地点的介词 in 省略；例 170 译文中引导地点的介词 at 省略。

3.4.3.2 省略某些介词搭配中的介词

例 171 Hydrogen is the lightest element with an atomic weight of 1.008.
【译文】氢是最轻的元素，原子量为 1.008。
例 172 The difference between the two machines consists in power.
【译文】这两台机器的差别在于功率不同。
例 173 The molecular structure is different for various kinds of polymers.
【译文】各种聚合物的分子结构不同。
例 174 Most of the patients with neuron circulatory asthenia are unwilling to accept psychotherapy as such.
【译文】多数神经循环无力患者不愿接受这种精神疗法。
例 175 It is necessary to develop electric power industry at a high speed.
【译文】必须加速发展电力工业。
例 171 译文中省略介词 with；例 172 译文中省略介词 between；例 173 译文中省略介词 for、of；例 174 译文中省译了介词 with；例 175 译文中省译了介词 at。

3.4.4　连词的省略

英语中的连词体系复杂、使用繁多，词与词之间、句子与句子之间的关系也主要由连词表示。此外，英语中无论是并列关系的连词还是主从关系的连词在使用时的语法结构比汉语稳定得多，而汉语词与词之间、句子与句子之间的关系常常靠词序来体现，连词的使用也多属于习惯或语气问题。因此，在翻译有连词的英语句子时，应视不同情况灵活处理，不要死抠原文。在很多情况下，英语的连词是可以省略不译的。省译连词的目的是使译文文字简洁，避免语意含混和词语重叠。

3.4.4.1　省略并列连词

例 176　This method is simpler and more direct than the previous one.

【译文】这种方法比前一种更简单直接。

译文中 and 根据汉语习惯省略不译。

例 177　Therefore the transformer can be used to step up or step down alternating voltage.

【译文】因此，变压器可用来升降电压。

译文中并列连词 or 可以省略不译。

例 178　Practically all substances expand when heated and contract when cooled.

【译文】几乎所有的物质都热胀冷缩。

译文中并列连词 and 可以省略不译。

3.4.4.2　省略从属连词

例 179　Ductility of a metal is usually much less when hot than when cold.

【译文】热状态的金属韧性一般要比冷状态的金属低得多。

例 180　As the temperature increases, the volume of water becomes greater.

【译文】温度升高，水的体积就增大。

例 181　A machine rotor has kinetic energy while it is set in motion.

【译文】正在运转的电机转子具有动能。

例 179 ~ 例 181 中时间状语的连接词 when，as 和 while 根据汉语习惯省略不译。

例 182　It is of course true that the working stresses should be below the elastic limit of the material.

【译文】材料的工作应力应低于其弹性极限。

例 183　Note that it is an e.m.f. that is generated and not a current.

【译文】注意，产生的是电动势而不是电流。

例 182 中主语从句的连接词 that 和例 183 中宾语从句的连接词 that 在汉译时省略。

3.4.5　动词的省略

英语句子多为主谓结构，而汉语句子中可以没有主语，其谓语除动词外，还可以直接用名词、形容词或词组（主谓词组、联合词组）来担任，也可以省略谓语动词。所以，在很多情况下英语动词在翻译成汉语时要作必要的省略，使译文更符合汉语表达习惯。

例 184　This laser beam covers a very narrow range of frequencies.

【译文】这种激光束的频率范围很窄。

例 185 Some substances, such as glass, rubber, and the like, offer a very high resistance.

【译文】某些物质（如玻璃、橡胶等）的电阻极高。

例 184、例 185 中的动词 cover、offer 根据汉语习惯省略不译。

例 186 It is clear that solids expand and contract as liquids and gases do.

【译文】很显然，固体像液体和气体一样，也会膨胀和收缩。

例 187 Glass and rubber offer a very high resistance.

【译文】玻璃、橡胶的电阻极高。

例 186、例 187 中分别省略了对重复动词 do 和动词 offer 的翻译。

除行为动词可以省略以外，系动词 be、become、get 等也可以省略。

例 188 The result of the experiment is rather satisfactory.

【译文】实验结果相当令人满意。

例 189 When the air pressure gets low, the boiling-point of water becomes low.

【译文】气压低，水的沸点就低。

例 188、例 189 中的系动词 is、get 和 become 在译文中被省略。

3.4.6 同义词或近义词的省略

英语中，有时某些词会与其同义词或近义词并列使用，或表示强调，或表示同一事物的不同名称。这样的句子在译成汉语时，应视情况采用省略译法。

例 190 The bonds in this crystal are of shared electron or covalent type.

【译文】这种晶体的化学键是共价键。

此例中 shared electron 和 covalent type 都是“共价键”的意思，译文中只译出一个，省略另一个。

例 191 Some hydraulic turbine generators are provided with amortisseur windings or damper windings.

【译文】某些水轮发电机具有阻尼绕组。

此例中 amortisseur windings 与 damper windings 表达同一意思“阻尼绕组”，译出一个，省译另一个。

例 192 Obviously, GA not only had a distinctly preventive effect on boll-shedding but also promoted noticeably the development of the boll.

【译文】GA 不仅具有防止落铃的显著作用，而且还能促进棉铃的生长发育。

原文中的 boll-shedding 和 the development of the boll 都表达“生长发育”的意思，译出一个，省略一个。

例 193 To be sure, the change of the earth is slow but, nevertheless, it is continuous.

【译文】地球的确变化很慢，但它确实是在不断地变化。

原文中的 but 和 nevertheless 表示同一个意思“但是、然而”，却被并列使用，这时也要译出一个，省略一个。

从以上各例句中不难看出，省译法不是任意省略、删减。某些词语如果不删减，译文必定不够简洁或累赘不堪，甚至影响整体意思的表达。采用省略法不仅使译文流畅、自然，而且突出了原文的整体意思。但是，应该特别注意的是省略绝不能影响原文的中心思想和整体意思，否则就是违背“忠实”这一基本原则。所以，采用省略法必须以不损害原文的内容为前提。此外，与增译法（amplification）相比，减省的情况要少得多，所以省略法要慎用，绝不能想删就删，随心所欲。

3.5　重　复

重复译法就是重复译出原文中出现过的某一词语或成分。英语和汉语一样，写文章总是尽量避免重复，然而在英译汉中重复却是一种必不可少的翻译技巧。这是因为翻译时往往需要重复原文中某些词，才能使译文表达更加明确具体。词的重复使用，在英汉两种语言里都常见，只是重复的词类和方式不同而已。比如，英语中常用代词、代动词重复出现过的名词或动词；而汉语则常常重复使用名词和动词。

重复法实际上也是一种增词法，只不过所增添的词是上文出现过的词。在科技英语翻译中，重复译法主要是为了表达上的明确具体，使译文句子结构匀称完整，其次是为了表达上的生动活泼。

3.5.1　代词的重复

3.5.1.1　人称代词的重复

英语中的人称代词 it 和 they 在翻译成汉语时常常利用重复法重复其所代替的名词。

例 194　Federally funded training and free-back-to-school programs for laid-off workers are under way, but few experts believe they will be able to keep up with the pace of the new technology.

【译文】为失业工人提供的联邦政府资助的培训计划和免费重返学校学习的计划目前都在实施中，但专家中几乎没有人认为这些计划能跟得上新技术的发展步伐。

例 195　If an atom contains three protons, it must have three electrons to be electrically neutral.

【译文】如果一个原子含有三个质子，这个原子必然有三个电子，使之不带电荷。

例 196　In China, they "made mistakes", suffered by them, acknowledged and studied them, thus planned victory.

在中国，他们“犯过错误”，吃过错误的亏，承认错误，研究错误，从而制定了胜利的方针。

例 194 中的 they 指的是上文中的 programs，在重复翻译时，要把代词所代替的名词译出来，而不是简单重复代词本身，这样才符合汉语的表达习惯；例 195 中的 it 应该重复译为“这个原子”；例 196 中的两个 them 指代前面提到的 mistakes（错误），在译文中被重复译出。

3.5.1.2　物主代词的重复

英语中用物主代词 its、his、their 等代替句子中作主语的名词时，翻译时往往可以不用代词而重复那个作主语的名词，以达到明确具体的目的。

例 197　The downward pressure exerted by water is proportional to its depth.

【译文】水向下所施加的压力和水的深度成正比。

例 198 The level of a liquid rises as its temperature is increased and falls with a decrease in temperature.

【译文】液面随着液体温度的升高而上升，随着温度的降低而下降。

例 197 中的 its 指的是 water，要译成“水的”；例 198 中的 its 则要译成“液体”。

3.5.1.3 指示代词的重复

英语中指示代词 that（those）、same 等较为多见，翻译时，仍然要重复译成它们所替代的名词。

例 199 The properties of alloys are much better than those of pure metals.

【译文】合金的性能比纯金属的性能要好得多。

例 200 We must check the conclusion in practice, and should not blindly rely on such as was reached merely by calculations.

【译文】我们必须在实践中验证这个结论，而不应轻易相信仅靠推算得出的这个结论。

例 201 Natural water is that which contains impurities.

【译文】天然水是含有杂质的水。

例 199 中 those 指代前面的 properties（特性），例 200 中 such 指代前面的 conclusion（结论），例 201 中的 that 指代前面提到的 natural water（天然水），在译文中均被重复译出。

3.5.1.4 不定代词的重复

不定代词的重译中，以 one 以及 so 等最为多见。翻译时，把 one 重复译成它所代替的名词，把 so 重复译成它所代替的句子或短语。

例 202 Other substances, apart from organic ones, burn in air or oxygen.

【译文】除有机物质外，其他物质在空气或氧气中也都燃烧。

例 203 An electric light bulb is a vacuum, and so is a radio tube.

【译文】电灯泡是真空的，电子管也是真空的。

例 204 A transversely stressed fillet weld can sustain higher loads than one stressed longitudinally.

【译文】受横向应力的角焊缝比受纵向应力的角焊缝承受能力强。

例 202 中 ones 指代前文中的 substances（物质），例 203 中 so 指代前文的 vacuum “真空”，例 204 中 one 指代主句中的 fillet weld（角焊缝），在译文中分别将它们重复译出，使中文意思更加明确。

3.5.1.5 代动词的重复

英语中的代动词 do 用途极广，常用来代替上文出现过的动词或动词短语，翻译时一般都需将其重复译为所代替的成分，以使表达鲜明具体。

例 205 It takes more power to do a job in two minutes than it does to do the same job in two hours.

【译文】两分钟做一件事所需的功率大于两小时做同一件事所需的功率。

原句中有三个 do，其中第一个和第三个 do 为实义动词，作“做”解；第二个 do（does）是代动词，代替了前文中的 takes。

例 206 Side forces in the vertical plane have a much smaller effect on stability than do horizontal side forces.

【译文】沿竖直面的侧向力对稳定性的影响，要比水平侧向力对稳定性的影响小得多。

原句中的 do 指代前文的 have。

例 207 Ferrous metals contain iron as their chief component, while non-ferrous metals do not.

【译文】黑色金属含铁为其主要成分，而有色金属则不含。

原句中的代动词 do 指代前面的 contain（含有），在译文中须译出。

3.5.2 多支共干结构中共有成分的重复

科技英语中存在这样的语言现象：几个动词共有一个宾语；几个宾语共用一个动词；几个形容词或介词（短语）共同修饰同一个名词；一个形容词或介词（短语）修饰几个名词，等等，这种现象被称为多支共干结构。

3.5.2.1 同词重译

所谓同词重译，指采用同一个汉语译词重复译出多支共干结构中的共有成分。

例 208 Gas, oil, and electric furnaces are most commonly used for heat treating metal.

【译文】金属热处理最常用的是煤气炉、油炉和电炉。

原句中 furnace 一词前有三个名词修饰语 gas，oil 和 electric，汉译时要分别译出才能使译文意思明确。

例 209 The chief effects of electric currents are the magnetic, heating, and chemical effects.

【译文】电流的主要效应是磁效应、热效应和化学效应。

原句中 magnetic、heating 和 chemical 三个形容词同时修饰 effects（效应），而在译文中须将“效应”一词分别加在“磁”、“热”和“化学”后面，使译文意思明确易懂。

例 210 Engineers have to analyse and solve design problems.

【译文】工程师要分析设计问题，解决设计问题。

原句中 problems 被 analyse 和 solve 两个动词所共有。

3.5.2.2 异词重译

异词重译指采用不同的汉语译词重复译出多支共干结构中的共有成分。

例 211 The levels of voltage, current and power are, on their own, not sufficient for demarcation.

【译文】仅凭电压的高低、电流的强弱和功率的大小尚不足以区别开来。

原句中的 levels 为其后三个名词 voltage、current、power 所共有，汉译时，依次用“高低”、“强弱”、“大小”三个词来修饰，进行重译。

例 212 You cannot build a ship, a house, or a machine tool if you do not know how to make a design or how to read it.

【译文】如果不会制图或是看不懂图纸，就不可能造出一条船，盖出一所房子，或生产出一台机床。

原句中 build 为其后三个名词 ship、house、machine tool 所共有，依次用“造出”、“盖出”、“生产出”三个词来修饰，进行重译。

例 213 He wanted to send them more aid, more weapons and a few more men.

【译文】他想给他们增加些援助，增添些武器、增派些人员。

原句中的动词 send 在译文中用“增加”、“增添”、“增派”三个不同形式的词重复译出。

3.5.3 修辞性重复

在实际翻译中，为了使译文的层次清楚，脉络分明，表达效果好，有时也可以把某些句子（无省略成分，也无多支共干结构）中的一些成分重复译出。这是汉语行文上的需要，涉及范围较广。

例 214 Every fuse has a definite current rating, indicating that no more than a specified number of amperes can pass through the fuse.

【译文】每一种保险丝都有一定的电流额定值，该额定值限定了保险丝所能通过的最大安培数。

例 215 There are a number of methods of joining metal articles together, depending on the type of metal and the strength of the joint which is required.

【译文】连接金属件的方法很多，选用哪种方法要视金属的种类和所要求的接点强度而定。

例 214、例 215 中分别重译了 current rating（额定值）、methods（方法）。

3.6 正反互译

汉语和英语在表达方式和表达习惯上存在着较大的差异，同样的意思，说法不尽相同。因此，在将英语翻译成汉语时，要根据具体情况采用正确的翻译方式，英语的正反译和反正译就是其中之一。顾名思义，“正反译”指的是把英文的肯定式译成汉语的否定式的翻译方法。比如，wonder 可译成“不知道”，difficult 可译成“不容易”，anything but 可译成“一点也不”。翻译时采用这一方法可使译文合乎汉语规范，恰当地表达原文的意思。“反正译”指的是把英文的否定式译成汉语的肯定式的翻译方法。英文中的 not ... until（直到……为止）结构和双重否定结构常采用这种方法来翻译。

在科技英语翻译中，究竟是采用正反译还是反正译，要根据上下文的具体情况做出选择。若有需要，各种词类的词语均可用词义的正反表达法来翻译。

3.6.1 正说反译

正说反译通常有两种情况：英语中某些单词或短语从形式上看是肯定的，但在意义上有明显的否定含义，此时宜采用正说反译法来翻译；当用“否定译肯定”比“肯定译肯定”更贴切时，也以正说反译为佳。

例 216 One wonders also why nature with some snakes concocted poison of such extreme potency.

【译文】人们也不明白，为什么大自然在某些蛇身上配制如此剧烈的毒液。

例 217　The performance of the machine is short of the requirements.

【译文】这台机器的性能没有达到要求。

例 218　The angularity of the parts is too great for proper assembly.

【译文】零件斜度太大，不宜装配。

例 219　In such cases an antifriction bearing might be a poor answer.

【译文】这时如果用减摩轴承，可能不是一种好的解决方法。

例 220　Indirect evidence suggests that Neptune's rings exist, but as arcs rather than true rings.

【译文】间接的证据表明海王星的光环是存在的，但它是弧形，而不是真正的环形。

例 221　Worm gear drives are quiet, vibration free, and extremely compact.

【译文】涡轮传动没有噪音，没有振动，而且非常紧凑。

例 222　Pure water hardly conducts an electric current at all, but it becomes a good conductor if salt is dissolved in it.

【译文】纯水几乎是不导电的，但是，如果其中溶解了盐就成了良导体。

例 223　These experimental values agreed with the theoretical values within the accuracy of ±0.1%.

【译文】这些实验数据与理论值相符，误差在±0.1%范围内。

例 224　We tried in vain to measure the voltage.

【译文】我们原想测量电压，但没测成。

例 225　Hardened steel is too hard and too brittle for many tools.

【译文】淬火钢太硬、太脆，许多工具不能用它制造。

例 226　Ideal machines which would have an efficiency of 100% should be free of friction.

【译文】效率为100%的理想机器必须没有摩擦。

例 227　The need for more potassium compounds than could be obtained from plants led men to search for other source of these important compounds.

【译文】从植物所能得到的钾化合物不能满足需要，这导致人们去寻找这些重要化合物的其他来源。

例 228　In the absence of force, a body will either remain at rest or continue to move with constant speed in a straight line.

【译文】没有外力，物体不是保持静止状态，就是做匀速直线运动。

例 229　The structure will prove weak in service.

【译文】使用中会证明该构件不牢固。

例 230　The precision instrument must be kept free from dust.

【译文】精密仪器必须保持无尘。

例 231　The Theory of Relativity worked out by Einstein is above many people's comprehension.

【译文】现在还有不少人不理解爱因斯坦提出的“相对论”。

原句中above表示品质、行为、能力超出某种范围之外，汉译时常用反说。

例 232　The specification lacks detail.

【译文】这份说明书不够详尽。

例 233 Better to do well than to say well.

【译文】与其说得好，不如做得好。

better... than... 这个结构是由 it is better... than to... 简化而来，than 后面所接的不定式表示被否定的对象，可译成"与其……不如……"。

例 234 The common gem materials tend to be less ductile and weaker.

【译文】一般的宝石材料不易延展，强度较差。

例 235 As rubber prevents electricity from passing through it, it is used as insulating material.

【译文】由于橡胶不导电，所以用作绝缘材料。

例 236 There are many other energy sources in store.

【译文】还有很多种其他能源尚未开发。

例 237 In the high altitude snow and ice remain all year.

【译文】海拔高的地方冰雪常年不化。

3.6.2 反说正译

反说正译与正说反译一样，可用于对英语句子中的名词、动词、形容词、副词和介词等词语的反译。英语中有些句子从形式上看是否定句，但实际上是肯定句，这时大多要反译。此外，英语中的双重否定经常可以采用以肯定译否定的方法来译，这是由于英汉两种语言表达习惯不同，以肯定译否定能更确切地表达原意。

例 238 It was suggested that such devices should be designed and produced without delay.

【译文】有人建议立即设计和生产这种装置。

例 239 Sodium is never found uncombined in nature.

【译文】自然界中的钠被发现都处于化合状态。

例 240 In this case we cannot but determine K first.

【译文】在这种情况下，我们只好先确定 K 的数值。

原句中 cannot but + 动词不定式，译为"不得不，只好"。

例 241 It's not easy to talk about Dolly in a world that doesn't share a uniform set of ethical values.

【译文】世界各地的伦理观念不同，因此就"多利"克隆羊的讨论难以达成一致意见。

例 242 Such flight couldn't long escape notice。

【译文】这类飞行迟早会被人发觉的。

例 243 Crystals do not melt until heated to a definite temperature.

【译文】晶体要加热到一定温度才会熔化。

not + 终止性动词（如 leave、come、go、receive、finish、stop 等）+ until...，意为"直到……才……"。

例 244 Laser is the most powerful drilling machine, because there is nothing on earth which cannot be drilled by it.

【译文】激光器是打孔能力最强的钻床，因为它能给地面上任何物体钻孔。

例 245　One body never exerts a force upon another without the second reacting against the first.

【译文】一个物体对另一个物体施加作用力必然会受到另一物体的反作用力。

例 246　If houses are at rest relative to the earth surface，the earth itself is not motionless.

【译文】如果房子相对于地球表面来说是静止的，地球本身却是运动的。

例 247　Nothing can be done about the external noise except change the geographical position of the receiver.

【译文】只有改变接收机的地理位置才能消除这种外部噪音。

例 248　There is nothing like mineral water to quench one's thirst.

【译文】矿泉水是解渴的最好的饮料。

所谓双重否定，就是否定之否定，同汉语的目的一样，是为了加以强调。不管形式如何，翻译时可直接译成肯定意义。

例 249　Despite its many advantages，wood are not without its drawbacks as a fuel.

【译文】尽管有很多优点，木材作为燃料仍有缺陷。

例 250　Dynamics is a discipline that cannot be mastered without extensive practice.

【译文】动力学是一门要做大量练习才能掌握的学科。

例 251　There is no rules that has no exception.

【译文】任何规则都无不例外。

例 252　There is not any advantage without disadvantage.

【译文】有一利必有一弊。

例 253　For practical reasons，it is impossible to maintain this figure in manufacturing without great cost.

【译文】由于实际原因，要保持这一生产数字，费用必然很高。

例 254　Hardly a month goes by without word of another survey revealing new depths of scientific illiteracy among U. S. citizens.

【译文】美国公民科学知识匮乏的现象日益严重，这种调查报告几乎月月都有。

例 255　The importance of proper lubrication cannot be overemphasized.

【译文】应特别强调适当进行润滑的重要性。

例 256　In the absence of electricity，large scale production is impossible.

【译文】大规模的生产必须靠电能。

综上所述，正反译和反正译是非常实用的英汉翻译技巧，能使译文更加简洁、明了，更符合汉语的表达习惯。但要注意的是，不管其适用范围有多广，也只能用来处理某一部分语言现象。英语中还有不少词，从正面和反面翻译都可以，译者应根据前后语境选用一种能更加贴近原文的表达方法。

3.7　否定句的翻译

英语和汉语都存在着大量的否定句，但由于历史、文化的不同，思维方式各异，反映在语言中的否定形式也不尽相同。一般说来，英语中表达否定的手段有：使用否定词；使用含有否定意义的词语；通过上下文或特殊句式体现否定。否定句是语法上的分类，它表示的是以否定词构成的句子形式。否定句的形式有部分否定、全部否定；有字面结构为肯定句，但实际含义为否定；有双重否定，但实际意义为强烈肯定。

总体来说，英语的否定词和表达方式有以下几种：全部否定（no、not、none、never、nothing、nobody、nowhere、neither、nor）；部分否定（not every、not all、not both、not much、not always）；几乎否定（hardly、scarcely、seldom、barely、few、little）；隐含否定含义的词汇（fail、without、beyond、until、unless、lest、ignorant、refrain、refuse、neglect、absence、instead of、other than、expect、rather than）。

3.7.1 全部否定句的译法

所谓全部否定是指否定整个句子的全部意思。全部否定句的否定词常用的有 no、not、nor、none、never、neither、nobody、nohow、nowhere、nothing 等。这些否定词不论在句中做什么成分，其所在的句子大都要译为全部否定。但需要把表示否定的“不、无、非”之类的词语用在动词前以构成全部否定。

例 257 He is not a mathematician.

【译文】他不是数学家。

not 和 no 是英语中最常用的两个否定词，二者有相同之处，也有不少区别。如果此句改为 He is no a mathematician，则译为“他根本不是数学家”。此时 no 的否定语气较强，有极端否定的意味，带有一定的感情色彩。而 not 的否定语气较弱，一般无感情色彩。

例 258 Gas has neither definite shape nor definite volume.

【译文】气体既没有一定的形状，也没有一定的体积。

例 259 Neither refractory materials nor magnetic materials can be used to make the part.

【译文】耐火材料和磁性材料都不用于生产这种零件。

例 260 Neither a positive charge, nor a negative the neutron has.

【译文】中子既不带正电荷，也不带负电荷。

例 261 Rubber is not a conductor, and neither is plastics.

【译文】橡胶不是导体，塑料也不是导体。

以上例句中的 neither... nor 是英语中用来表示全部否定意义的词，应将其完全直译出来，即“既没有……也没有”或“都不”。

例 262 None of these metals have conductivity higher than copper.

【译文】这些金属的导电率都不及铜高。

none 在原句中起到了完全否定的作用，因而翻译成“都不”。

例 263 None of the inert gases will combine with other substances to form compounds.

【译文】无论哪一种惰性气体，都不会和其他物质化合而形成化合物。

例 264 None of these substances are good conductors of electricity.

【译文】这些物质都不是电的良导体。

原句中 none 表示“没有一个”，是全部否定句，在译成汉语时要做一些修订，即否定的转移，但仍保持原句的全部否定含义，译为“都不是”。

例265 During ordinary chemical reactions the nucleus of an atom does not undergo any changes.

【译文】在一般的化学反应中，原子核不发生变化。

例266 At least nowadays, there is no way to harness the energy of fusion.

【译文】至少目前还没有办法利用聚变能。

例267 Nothing in the world moves faster than light.

【译文】世界上没有任何东西比光传得更快了。

例268 The effect of the medicine is such a mild one as not to stimulate the sweat glands.

【译文】这种药作用温和，不会刺激汗腺。

例269 Nowhere in nature is aluminum found free, owing to its always being combined with other elements, most commonly with oxygen.

【译文】由于铝总是和其他元素结合在一起，最常见的是和氧结合，因此在自然界根本找不到游离态的铝。

3.7.2 部分否定句的译法

所谓部分否定，就是对叙述的内容做部分的而不是全部的否定。有些不定代词（如all、everything、ever、each、both）及副词（如always、often、wholly）与否定词not连用时，表示的概念是否定整体中的一部分，而不是全部否定，相当于汉语的“不是所有都”、“不是每个都”、“不是两个都”、“不总是”之意，从形式上看很像全部否定，而实际却是部分否定，翻译时要特别注意。翻译成中文时应为“并非所有”、“并不都”等。

这种部分否定句一般可分为两种情况，一种是含有both的否定句；一种是含有all、every、many等的否定句。

3.7.2.1 含有both的部分否定句的译法

在含有both的否定句中，不管否定词位于什么位置上，都表示部分否定，both一定是否定的重点。通常译为“不都是”，“并非都”，“并不都”。

例270 Both of the instruments are not precision ones.

【译文】这两台仪器并非都精密。

例271 Both of the substances are not made up of carbons.

【译文】这两种物质并非都是由碳组成的。

例272 We are not familiar with both of the instruments.

【译文】我们对这两台机器不是都熟悉。

3.7.2.2 not在all、every、always等词之前的部分否定句的译法

当否定词not位于all、every、always等词之前时，表示部分否定，并且后者一定是否定的重点，不存在其他可能。通常译为“不全是”、“不是所有……都”、“并非都”等。

例273 Positive ions are not all alike and may differ in charge or weight.

【译文】正离子并不完全相同，他们在电量和重量上都可能有差别。

例274 An engine may not always do work at its rated horse-power.

【译文】发动机并不总是以额定马力工作。

例275 Not everyone can be a mathematician, but in order to understand our modern world，it is necessary to know something about mathematics.

【译文】并不是人人都能成为数学家，但是为了了解现代世界，有必要对数学有所了解。

例 276 Digital oscilloscopes can not be often used in our experiments.

【译文】数字示波器不常用在我们的实验中。

3.7.2.3 not 在 all、every、always 等词之后的部分否定句的译法

当否定词 not 不在 all 等词之前，而在句中谓语的位置时，则有两种可能：部分否定或全部否定。这类句型究竟是按部分否定译出或是按全部否定译出，应根据上下文和具体的语境而定。

例 277 All metals do not conduct electricity equally well.

【译文】并非所有金属的导电性能都同样好。

例 278 All these various losses，great as they are，do not in any way contradict the law of conservation of energy.

【译文 1】所有这些各种各样的损失，虽然都很大，却并不都是和能量守恒定律矛盾的。

【译文 2】所有这些各种各样的损失，虽然很大，却都和能量守恒定律没有一点矛盾。

译文 1 是按照部分否定来译的，译文 2 是按照全部否定来译的，两种译文在语法上都是合理的。但根据概念和常识判断，按全部否定译出的译文是正确的。也就是说，all... not 在这里表示全部否定。

例 279 All the chemical energy of the fuel is not converted into heat.

【译文 1】并非燃料中的所有化学能都可转化为热。

【译文 2】燃料中的所有化学能都不能转化为热。

译文 1 是按照部分否定来译的，译文 2 是按照全部否定来译的，两种译文的意思相去甚远。就单个句子看，在语法、概念、逻辑上都无错误。但是根据概念和常识，按部分否定译出的译文更合乎情理。也就是说，all... not 在这里表示部分否定。

例 280 All isotopes cannot be manufactured in this way.

【译文】并非所有的同位素都可以这样制造。

3.7.3 几乎否定句的译法

几乎否定句又称准否定句或半否定句。英语中有些表示否定的词语，意思上接近 never、not、no、none 等词，这类词叫几乎否定词，由几乎否定词构成的句子叫几乎否定句。常用的表示几乎否定的词有：few、little、hardly、seldom、scarcely、barely、almost no 等。翻译成汉语时，可译为“极少”、“几乎没有”、“几乎不”、“很少”等。这类否定词在表示否定意义时不像 no、not、never 等词那么绝对，而是留有余地，语气也比较弱。

3.7.3.1 几乎否定副词的译法

几乎否定副词有 hardly、scarcely、barely、rarely、seldom、scantly、infrequently、unfrequently、uncommonly、only、little 等，在汉译时可直译。

例 281 Curiously，Uranus has almost no such excess heat.

【译文】奇怪的是，天王星几乎没有这种多余的热量。

例 282 Mercury, so small and close to the sun that its gases were quickly lost to space, is nearly airless.

【译文】水星上面几乎没有空气，因为它体积很小，并且离太阳很近，所以它的气体很快就散失在太空中。

例 283 The speed of the man-made satellite hardly changes at all.

【译文】这颗人造卫星的速度几乎没有什么变化。

例 284 Rarely do metals occur in nature in a pure form by themselves.

【译文】在自然界，金属很少以纯金属的形式存在。

例 285 The US has well-developed and successful offensive command and control warfare (C2W), electronic warfare (EW), and other information warfare (IW) capacities, but these can hardly be characterized as "strategic".

【译文】美国拥有健全和成功的进攻性指挥与控制战（C2W）、电子战（EW）和其他信息战（IW）的能力，但这些几乎不能被称做“战略性的”。

例 286 Barely any of our present batteries would be satisfactory enough to drive the electric train fast and at a reasonable cost.

【译文】我们现有的蓄电池几乎都不足以保证电气火车快速而经济高效的运行。

例 287 Scarcely ever does the common oyster contain a valuable pearl.

【译文】普通牡蛎中难得含有有价值的珍珠。

几乎否定副词提到句首时，句子语序通常倒装。但修饰主语时例外。

例 288 Scarcely any enemy planes were left undamaged after the guerrilla attack.

【译文】在游击队袭击之后，敌人几乎没有一架完好的飞机。

3.7.3.2 几乎否定代词的译法

几乎否定代词和形容词有 little，few 等，译成汉语时，仍然采用否定形式。

例 289 The earliest forms of jet-propulsion had little ability to function at rest, in view of the absence of any means of air-compression.

【译文】最早的喷气发动机由于缺乏空气压缩手段，静止时几乎没有工作能力。

例 290 Few believed the rumor.

【译文】相信这个谣言的人不多。

例 291 Because the body's defense system is damaged, the patient has little ability to fight off many other diseases.

【译文】由于人体的免疫系统遭到破坏，病人几乎没有什么能力来抵抗许许多多其他疾病的侵袭。

例 292 There is little fuel left in the tank.

【译文】油罐里几乎没有什么燃油。

3.7.3.3 含有几乎否定词的短语的译法

含有几乎否定词的短语有 nearly nothing、nearly no、scarcely any、hardly ever、little or nothing、seldom 等，可译为“几乎……”。

例 293 Gorilla gained scarcely anything.

【译文】大猩猩几乎什么也没得到。

例 294 Unfortunately little or nothing is left of the former splendour.

【译文】可惜，昔日的壮观已经所剩无几或踪迹全无了。

例 295 There are few, if any, such situation.

【译文】这种情况，即使有，也很少。

3.7.3.4 含有几乎否定意义的短语的译法

短语 almost no、almost never、next to、next door to 等常含有几乎否定意义，译成汉语时经常为“几乎 + 否定词”的形式。

例 296 There seems almost no difference between these two kinds of alloys.

【译文】这两种合金看上去几乎没有什么差别。

例 297 It's next to useless to simply prove hypothesis in theory.

【译文】仅从理论上证明这一假设，几乎毫无价值。

例 298 In those days when science remained undeveloped, the crew's knowledge of thunder and lightening was next to nothing.

【译文】在那科学不发达的年代里，水手们对于雷电几乎一窍不通。

3.7.4 双重否定句的译法

双重否定注重修辞作用，是由两个否定意义的词在同一个句子里出现而构成双重否定的语言现象，即否定之否定。在翻译时，既可直译成双重否定，也可转译成肯定。在译文中如果保留否定的语气，比单纯的肯定强而有力。双重否定是一种生动而强调的肯定，其形式变化颇多，通常是由 not、no、never、neither、nobody、nothing 等否定词和其他含有否定意义的词或词组搭配而构成的。该结构常被译为“没有……就不……”或“没有……就没有……”。

双重否定句的主要类型包括：

否定词 + 含否定词缀（un-、in-、dis-、non-、-less 等）的词；

否定词 + without + 名词（动名词）；

否定词 + 否定意义的词（如 keep from、refuse、neglect、forget 等）；

主句（否定结构） + 从句（否定结构）；

否定词 + but（but 可作介词，连词，关系代词）。

最常见的双重否定形式有：no（not）... no（not）...（没有……没有……）、no（not）... but...（没有……不……）、no（not）... without...（除……不……）、not（none）... the less...（不因……就不……）、no（not）... unless...（除非……才……）、not... until...（直到……才……）、no（none）... other than...（不是别的……而是……）、not but that...（并非……不……）、not a little...（不少；很多；大大地）。

3.7.4.1 直译为双重否定句

例 299 In short, microwaves can only travel in straight line. Without a series of relay towers, it is impossible for them to send message over long distances to remote places.

【译文】总之，微波只能以直线方式传播，要是没有一系列的转播塔，微波就不能越过漫长的距离，将消息传递到远方。

例 300 There is no way the shuttle could ever be operated as a purely commercial vehicle without any National Aeronautics and Space Administration (NASA) funding.

【译文】如果没有美国宇航局的任何资助，航天飞机根本不可能作为一种纯商业航天器运行。

例 301 If it were not enough acceleration, the earth satellite would not get into space.

【译文】要是没有足够的加速度，地球卫星是不能进入太空的。

例 302 Because heat does not take up any room and it does not weigh anything, it is not a material.

【译文】热不占有任何空间，也不具有什么重量，因此它不是物质。

例 303 In the absence of radar, the pilot in an airplane could not fly for a long distance at night.

【译文】没有雷达，飞行员在夜间不能进行长途飞行。

介词短语 in the absence of 在句中译为“没有”，经常见于双重否定句中。

例 304 But for the sun, all living things could not live on the earth.

【译文】要是没有太阳，所有生物在地球上都不能生存。

but for + 名词表示“要不是”和“没有”之意，一般用于虚拟条件句中。

3.7.4.2 转译为肯定句

例 305 There is no modern communication means that has no disadvantage.

【译文】现代通讯手段都有缺点。

原句中的两个 no 构成了双重否定句，但根据句子的含义，可以翻译成汉语的肯定句，因而译为“都有”。

例 306 Now, no spaceship cannot be loaded with man.

【译文】现在的宇宙飞船都能够载人了。

原句中的 no 与 cannot 构成双重否定，但本句可以按汉语的习惯译为肯定形式“都能够”。

例 307 The flowing of electricity through a wire is not unlike that of water through a pipe.

【译文】电流过导线就像水流过管子一样。

例 308 There can be no sunshine without shadow.

【译文】有阳光就有阴影。

例 309 One can tell the difference almost at a glance, for a spider always has eight legs and an insect never more than six.

【译文】人们一眼就能注意到蜘蛛与昆虫的区别，因为蜘蛛总是有八条腿，而昆虫的腿只有六条。

原句中的 never more than 在此处译为“只有，仅仅”。

3.7.5 否定的转移

否定转移是指否定形式在谓语动词，而否定的信息焦点却在状语和表语，或否定形式在主句，而否定的信息焦点却在从句。英语的否定词在句中究竟否定哪个部分，对于理解全句的语义至关重要。这是英语的一种习惯思维方法，这类否定句型的翻译必须按照汉语表达习惯进行否定成分的转译，这样才能避免生搬硬套、文理不通的现象。

3.7.5.1 否定主语转换为否定谓语

例 310 The experiment on the transformation of energy shows that no energy can be created and destroyed.

【译文】能量转换的实验说明了能量既不能创造，也不能毁灭。

原句中的否定词 no 是用来否定从句中的主语 energy 的，但在译文中却用来否定两个谓语动词“创造”和“毁灭”。

例 311 In the printing industry there are very few parts that have a predetermined life expectancy.

【译文】在印刷业中，任何零部件几乎都无法预测其生命期限。

原句中的 very few 是否定名词的，翻译成汉语时却用来否定动词。

3.7.5.2 否定状语转换为否定谓语

例 312 The Yellow River goes by Jinan, not through the city.

【译文】黄河流经济南市郊而不穿过市区。

例 313 Nowhere in nature is aluminum found free.

【译文】在自然界任何地方都找不到游离状态的铝。

Nowhere 在原句中作主语，在译文中作谓语。

3.7.5.3 否定宾语转换为否定谓语

例 314 We know of no effective way to store solar energy.

【译文】我们不知道储存太阳能的有效方法。

例 315 Snakes are different from the other reptiles because their bodies are very long and they have no legs.

【译文】蛇与其他爬行动物不同，因为蛇的身体很长，而且没有足。

3.7.5.4 否定介词宾语转换为否定谓语

这些表示否定意义的介词宾语包括 on no conditions、under no circumstances、in no circumstances等。

例 316 Workers can violate the safety rules on no conditions.

【译文】工人们在任何情况下都不能违反安全规则。

例 317 We shall consent to the designing plan under no circumstances.

【译文】在任何情况下，我们都不会同意这种设计方案。

3.7.5.5 否定主句转换为否定从句

英语原句否定主句，翻译成汉语的时候，需要把否定转移到从句上。常见的结构是 not... because...，可以翻译为“并不是因为……才……”。

例 318 He was not ready to believe something just because Aristotle said so.

【译文】他并不只是因为亚里士多德说过某事如何如何，就轻易相信它。

例 319 The engine did not stop because the fuel was finished.

【译文】发动机停了，不是因为燃料用完了。

3.7.5.6 否定谓语转换为否定补语

例 320 This electric motor does not work properly.

【译文】这台电动机运转得不正常。

例321 If iron is kept in air-free distilled water, its rusting is not so fast.

【译文】如果铁放在无空气的蒸馏水中，它生锈得并不快。

3.7.5.7 否定谓语转换为否定状语

例322 He doesn't study in this institute.

【译文】他不在这所学院读书。

例323 Metals do not change their forms as easily as plastic bodies do.

【译文】金属不像塑料物体那样容易变形。

3.7.5.8 否定主句的谓语转换为否定宾语从句的谓语

这类否定的转移常常出现在动词 think、believe、expect、suppose、seem、feel、look、imagine、reckon、fancy、anticipate、figure 等作谓语且后接宾语从句或不定式短语的否定句中。翻译时需要把否定转移到宾语从句的谓语动词前面或不定式短语前面。

例324 I don't suppose they will object to my suggestion.

【译文】我想他们不会反对我的建议的。

例325 For many years the atom was not believed to be divisible.

【译文】多年来原子一直被认为是不可分割的。

原句中的谓语部分使用了动词 believe，并且其后接不定式短语，否定成分在译成汉语时要转移。译文中的“不可分割”即是转移后的译文，符合汉语的表达习惯。

3.7.6 意义否定句的译法

意义否定句也称含蓄否定句或内容否定句。英语里有些句子形式上是肯定句，但实际上是否定句。关于这种意义否定句的翻译，可参阅上一节所论述的“正说反译”部分的有关例句。以下分类阐述意义否定句。

3.7.6.1 名词引起的否定

含有名词 neglect（忽视）、absence（缺失）、loss、exclusion（排除）、refusal（拒绝）、failure（失败）、deficiency（不足；缺乏）等的句子，其句子内容是否定的。

例326 A few instruments are in a state of neglect.

【译文】一些仪器出于无人管理状态。

例327 And in true chicken-or-the-egg fashion, the lack of course ware deters the purchase of computers by schools.

【译文】这真是一个先有鸡还是先有蛋的问题，没有教学软件，学校就不敢购买计算机。

3.7.6.2 动词或动词词组引起的否定

含有动词 deny（拒绝；否认）、exclude（排除；除去）、avoid（避免）、fail（不能；没有）、ignore（不顾；忽视）、lack（不足）、miss（失败；没有）、overlook（忽视；不顾）、stop（停止）/keep / prevent / protect / save... from（避免……；不受……）、refuse（拒绝）、neglect（忽略）等的句子，其句子内容是否定的。

例328 They shut off valves to prevent steam from reaching the turbine they intend to test.

【译文】他们关上阀门不让蒸汽流到他们要实验的涡轮。

例329 According to preliminary calculation, it costs a total of some $800, on average, to stop a ton of sulphur from getting into the air.

【译文】根据初步计算，阻止一吨硫进入空气平均要花费约800美元。

3.7.6.3 形容词或形容词短语引起的否定

含有形容词及形容词短语 little、few、devoid of（没有）、ignorant of（不知道）、few（几乎没有）、short of（缺少）、far from（绝不）、little（几乎没有）、safe from（免于）、last（决不会）、free from（没有；免于）等的句子，其句子内容是否定的，因此，译成汉语的否定句。

例 330 The study confirmed that few people are likely to escape the effects of nuclear exchange.

【译文】研究证明几乎无人能逃脱互投核弹造成的恶果。

3.7.6.4 介词或介词短语引起的否定

含有介词及介词短语 beyond、above、beneath（不值得）、besides、out of（缺乏）、instead of（而不是）等的句子表达否定意义，可直译成汉语的否定句。

例 331 The problem is beyond the reach of my understanding.

【译文】这个问题我无法理解。

例 332 A computer may use several thousand tubes and transistors. But how a computer uses its tubes may seem out of understanding.

【译文】一台计算机可能使用几千只电子管和晶体管，但是计算机怎样使用它的电子管似乎是无法理解的。

3.7.6.5 连词及某些短语引起的否定

常见的引起否定的连词及短语有 before、but、would rather... than、rather than、too... to、far from 等，译成汉语的否定式。

例 333 Never start to do the experiment before you have checked the meter.

【译文】没有检查好仪表，切勿开始做实验。

例 334 They would rather use machine parts of plastics than use those of metals.

【译文】他们宁可使用塑料机器零件，也不用金属机器零件。

还有一些常用的结构表示否定意义，如 more than can... 在意义上相当于英语的 can not...，可以翻译为“简直不、无法、难以”；而 more than one can help 相当于 as little as possible，可以翻译为“尽量不、绝对不”；anything but...，常常翻译为“绝对不、根本不、一点也不”；have yet to do...，相当于 have not yet done...，常常翻译为“还没有”；may（might）as well... 结构常常翻译为“还不如”。

3.7.7 形式否定句的译法

英语句子中有些否定词与其他词连用形成一种固定搭配，表面看似乎是否定的，但实际所表达的含义却是肯定的，这种结构就是形式否定。关于这种形式否定句的翻译，可参阅上一节所论述的“反说正译”部分的有关例句。常见的带有隐含肯定意义的词组或单词主要有：not... until（直到……才）、not... too（越……越好）、none but（只有）、nothing but（只有，只不过）、nothing more than（仅仅）、no sooner... than（刚一……就）、none other than（不是别的人或物而正是）、none the less（依然，仍然）、not but that（虽然）、

make nothing of（对……等闲视之）、for nothing（徒然，免费）、not only...but also（不仅……而且）、not...long before（很快就）、no more than（和……一样不）、no other than（只有，正是）。以及 cannot...too、no...little、not...slightly、not...other than（可分别译为“无论怎样……都不过分”、“很多”、“就/只能……”）等。

例 335 It cannot be too much emphasized that agriculture is the foundation of the national economy.

【译文】农业是国民经济的基础，怎么强调也不过分。

原句中的 cannot...too 是英语中比较常见的否定形式表达肯定含义，翻译成汉语时要注意其强调性，此处可分别译为“无论如何……也不过分”。

例 336 The importance of the research project on automatic controls cannot be overestimated.

【译文】那项有关自动控制的科研项目的重要性，怎么估计也不过分。

例 337 The importance of the maintenance of the highest degree of sanitation and quality control in the operation of a commercially sterile cold fill system cannot be overemphasized.

【译文】保持高度卫生的重要性和商业上无菌冷装系统操作方面的质量控制，我们怎么强调也不过分。

例 338 The technological importance of temperature that marks discontinuities in the mechanical properties of materials can hardly be exaggerated.

【译文】材料力学性能上标志明显突变的温度，在工艺上的重要性怎么说也不过分。

例 336 中 cannot be overestimated 译为“怎么估计也不过分”；例 337 中 cannot be overemphasized 译为“怎么强调也不过分”；例 338 中 can hardly be exaggerated 译为“怎么（夸大）说也不过分”。

例 339 We cannot help admiring Madame Curie for her great achievements in scientific research.

【译文】我们对居里夫人在科学研究方面所取得的成就不胜钦佩。

此句也可改为 We cannot but admire Madame Curie for her great achievements in scientific research. cannot help + doing 与 cannot but + do 意思相同，都是“不禁、忍不住、不得不”之意。意思相似的表达方式还有 cannot keep（abstain 节制、refrain 抑制）from + doing。

例 340 The star is no brighter than that one.

【译文】这颗恒星和那些恒星一样不亮。

或译：这颗恒星并不比那颗恒星更亮。

或译：这颗恒星和那颗恒星一样暗淡。

no more than 表示 than 前后两个事物程度相同，而且都是否定的，其意义等同于 not...any more than，通常可译为“和……一样不”、“并不比……更”。

3.7.8 延续否定句的译法

延续否定句指的是前面已经有一个否定句，为了进一步说明前面的否定概念，后面又追加一个或数个否定句，这种追加的否定句叫做延续否定句。延续否定是一种强调的否定，一个否定概念在句中多次重复（用相同方式或不同方式），其目的是为了更明确地表达否定意思。它既不是双重否定，也不是重复否定。在否定句后加一些具体的内容加以否定，以补充强调前面的否定。翻译时应该用适当的汉语词语表达出同等程度的强调，如使用“决不”、“更不用说”等。延续否定句有以下几种表达形式：

3.7.8.1 一个或多个意义相近的否定词在句中重复出现

例 341 We don't retreat, we never have, and never will.

【译文】我们不后退，我们从未后退过，将来也决不后退。

例 342 It was not anger, nor surprise, nor disapproval, nor horror, nor any of the emotions she had been prepared for.

【译文】那表情不是生气、不是惊讶、不是不满、不是厌恶，也不是她所估计的任何一种感情。

例 343 A woman was distributing a brochure that encouraged readers not to use anything that comes from or is tested in animals——no meat, no fur, no medicines

【译文】一位妇女在散发小册子，鼓励读者不要使用任何来自动物身上的东西，不要食用动物的肉，不要使用动物的皮毛，不要用动物做药材。

例 344 The elephant isn't like a wall, or a spear, or a snake, or a tree, neither is he like a fan。

【译文】这只象不像一堵墙，不像长矛，不像蛇，不像树也不像扇子。

例 345 The moon had no seas, lakes or rivers or water in any form. There are no forests, prairies or green fields and certainly no towns or cities.

【译文】月球上没有海、湖或者河，或者任何形式的水，没有森林、草原或者绿色田野，肯定没有城镇或城市。

原句重复使用 or 引出多个追加的否定内容。or 在肯定句中一般可译为“或者”，但是在否定句中相当于 and not，因此应译为“没有”。

3.7.8.2 否定句 + to say nothing of (let alone/not to speak of) + 对称成分

例 346 In old China there was hardly any machine-building industry, not to speak of an aviation industry.

【译文】在旧中国几乎没有什么机器制造工业，更不必说航空工业了。

例 347 Three people were badly hurt, to say nothing of damage to the building.

【译文】三个人受了重伤，更不用说房子遭受的损害。

3.7.8.3 两个否定句否定同一个对象

例 348 None of scientists doesn't know what happened to it, no one.

【译文】没有一个科学家不知道它发生了什么，没有谁（不知道）。

例 349 You don't know the of operation of the machine, neither/nor do I.

【译文】你并不了解机器的运行原理，我也不了解。

3.7.8.4 否定句 + 延续否定词 + 谓语

例 350 The remote server does not support encryption and secure connection.

【译文】远程服务器不支持加密，也不支持安全连接。

例 351 This structure could never have been brought into being by following overall design or rational overall arrangement.

【译文】这个建筑是不可能有总体规划设计的，也不可能有合理的布局。

例 352　Act finding is a difficult problem. It may be said that some of the facts could never be clarified even by Confucius or Aristotle.

【译文】寻找事实是一件很难的事情，可以说有一些事实是根本弄不清的，即使是孔子、亚里士多德在世，他们也不一定能搞清。

例 353　Never covet wealth and power.

【译文】切勿贪图财富和权力。

例 354　The purpose of this chapter is not to describe the various fasteners or tabulate available sizes.

【译文】本章不是要讲述各种紧固件，也没有用表格列出可用的各种尺寸。

翻译练习

1. The thickness of a tooth measured along the pitch circle is one half the circular pitch.
2. The earthquake measured 6. 5 on the Richter scale.
3. We must reflect what measures to take in case of any accidental collapse of a bed.
4. In the transistor the output current depends upon the input current, hence it is a current-operated device.
5. The relay is operated by a current of several milliamperes.
6. The hearing aids are operated from batteries.
7. Either of two reactions may be in effect in the reduction of iron oxide with carbon.
8. Alloys belong to a half-way house between mixtures and compounds.
9. The spindle rotates simultaneously round two axes at right angles to each other.
10. An electron is an extremely small corpuscle with negative charge which rounds about the nucleus of an atom.
11. The rudder serves the purpose of yawing the airplane to the right or left.
12. Each of these compounds boils at a different temperature.
13. After more experiments, Galileo succeeded in making a much better telescope.
14. During the two and half hour talk, the two sides exchanged views on the choice of terms of payment, but they made no mention of the mode of transportation.
15. The application of electronic computers makes for a tremendous rise in labor productivity.
16. The maiden voyage of the newly-built steamship was a success.
17. About 20 kilometers thick, this giant umbrella is made up of a layer of ozone gas.
18. We have made a careful study of the properties of these chemical elements.
19. Gases and liquid are perfectly plastic.
20. That radio factory impressed me deeply.
21. All of the austenitic stainless steels resist hydrogen damage.
22. If the N pole of one magnet is held toward the N pole of another, the magnets will push away from each other.
23. Simple in principle, the experiment led to a scientific revolution with far-reaching consequences.
24. A list of all the ways that diesel power is used would take pages.
25. The concentration of carbon dioxide in the air being only 0. 03 per cent, carbon is amassed into the compass of the plant from a large volume of air.

26. Ice and water consist of the same substance in different forms.

27. Were there no gravity, there would be no air around the earth.

28. A scientist constantly tried to defeat his hypotheses, his theories, and his conclusion.

29. Solids transmit sound very well.

30. This typewriter is indeed cheap and fine.

31. James Watt invented the steam engine.

32. I can finish the work so long as you give me time.

33. If you don't bring the map, you'll get lost.

34. We should gradually eliminate the differences between town and country.

35. Applicants who have worked at a job will receive preference over those who have not.

36. He put his hands into his pockets and then shrugged his shoulders.

37. Visible light covers a wave length range of about 0.38 to 0.78 μm.

38. When the masses are of one heart, everything becomes easy.

39. It is not so easy to get iron from its ore.

40. It's no use quarrelling now.

41. It is the U. S. that is distorting and perverting the "Geneva spirit".

42. We eat to live, but not live to eat.

43. The industrial waste gases are harmful to us and we should by all means remove them.

44. One must make painstaking efforts before one could succeed in mastering a foreign language.

45. A wise man will not make such a mistake.

46. The earth goes around the sun.

47. The population of Guangdong Province is even larger than that of the U. K.

48. If you know the frequency, you can find the wavelength.

49. The jammer covers an operating frequency range from 20 ~ 500MHz.

50. Aluminum alloys can be divided into two classes: heat-treatable and non-heat treatable alloys.

51. All the metals are good conductors because there is a great number of free electrons in them.

52. Electrical inventors who followed Edison did not have to experiment with the substances which he had found would not work.

53. Basically, there are two directions of applied loads on the fillet welds: parallel to the weld length and transverse to the weld length.

54. Trusses may be classified as simple, compound, or complex.

55. Work does not include time, but power does.

56. We had meant to the laboratory, but we forgot to.

57. We have to analyze and solve problems.

58. Marketing economy is itself the product of long course of development, of a series of revolutions in the modes of production and of exchange.

59. Among the four pictures, two appear to be real, others false.

60. Actually, it isn't, because it assumes that there is an agreed account of human rights, which is something the world does not have.

61. The people of China have always been courageous enough to probe into things, to make inventions and to make revolution.

62. Our policy must be made known not only to the leaders and to the cadres but also to the broad masses.

63. Happy families also had their own troubles.

64. He was proficient both as a flyer and as a navigator.

65. We must actively introduce new techniques, equipment, technologies and material.

66. I had experienced oxygen and /or engine trouble.

67. Under ordinary conditions of pressure, water becomes ice at 0 °C and steam at 100 °C.

68. What we want, first and foremost, is to learn, to learn and learn.

69. Courage in excess becomes foolhardiness, affection weakness, thrift avarice.

70. We should discard the idea that scientific inquiry will ever be complete.

71. Ice is not as dense as water and it therefore floats.

72. Metals do not melt until heated to a definite temperature.

73. Until recently geneticists were not interested in particular genes.

74. We cannot be too careful in doing experiment.

75. Discovering that end product differs significantly from the prototype that was tested is not unusual.

76. The average speed of all molecules remains the same so long as the temperature is constant.

77. The influence of temperature on the conductivity of metals is slight.

78. The common non-metals tend to be less strong than metals.

79. The absence of air also explains why the stars do not seem to twinkle in space, as they do from the earth.

80. Eclipses can not be often seen in every part of the world.

81. There exist neither perfect insulations nor perfect conductors.

82. In contradiction to solids, gases never have a definite volume.

83. The absolute zero of temperature can never be reached.

84. IBM will probably never be able to regain its position as the undisputed colossus of the computer industry.

85. Horsepower has nothing to do with the horse.

86. Hardened steel and brittle materials such as glass and ceramics are not normally amenable to diamond turning.

87. In contradiction to solids, gases never have a definite volume.

88. Not everybody is convinced the Leaning Tower of Pisa really can be saved.

89. Not many of the things are of use in form in which they are found.

第 4 章　词语的翻译

英语词汇大致分为两大类：实词（Notional Word or Content Word）和虚词（Form Word，Functional Word or Empty Word）。实词也称为实义词，指本身具有完整词汇意义或语法功能的词。这类词汇可随社会、科学、技术的发展而变化。语言学家把这类词称为开放类词（Open Class Word）。这类词在句子中可独立担任成分，起着决定句子意义的关键作用，包括名词、动词、形容词、副词、代词、数词。虚词也称做功能词或形式词，指不能在句子中独立担任任何成分的词，它们必须同其他词类结合形成词组，才能在句子中充当成分。语言学家把这类词称为封闭类词（Close Class Word）。这类词在其发展过程中变化甚少，基本处于不变的情况。这类词通常在句子中占次要地位，包括介词、连词、冠词、感叹词。本章主要针对科技英语中一些重要的实词和虚词的翻译进行说明。

4.1　形容词的翻译

英语的形容词在句子中可以用作定语、表语、宾语补足语或主语补足语等。作定语的形容词一般放在它所修饰的名词之前，称之为前置形容词或前置定语；有时作定语的形容词也会放在所修饰名词之后，称为后置形容词或后置定语。

4.1.1　形容词的一般译法

4.1.1.1　直译

绝大部分前置形容词和一部分后置形容词可直译为形容词，构成所修饰名词的定语。

例 1　It is a hard and brittle material.

【译文】这是一种硬而脆的材料。

例 2　The medium carbon steel has a high melting point.

【译文】中碳钢的熔点高。

4.1.1.2　转译

（1）转译为名词

英语中有些形容词加上定冠词 the 表示某一类人或物，汉译时常译成名词。另外，有些表示事物特征的形容词作表语，在汉译时往往可在其后加上“性”、“度”、“体”等词，使其转译为名词。

例 3　In fission processes the fission fragments are very radioactive.

【译文】在裂变过程中，裂变碎片的放射性很强。

例 4　The cutting tool must be strong，tough，hard，and wear resistant.

【译文】刀具必须有足够的强度、硬度、韧性和耐磨性。

例 3 中作表语的形容词 radioactive 在汉译时转译为名词“放射性”；例 4 中的形容词

strong、tough、hard、wear resistant 均被译为相应的名词“强度”、“硬度”、“韧性”和“耐磨性”。

（2）转译为动词

英语中形容词在句中作表语、主语补足语或其他成分，或者形容词与介词搭配作表语或定语，在汉译时通常可译为汉语的动词。

例 5 They are very familiar with the properties of this type of thermoplastics.

【译文】他们十分熟悉这类热塑性塑料的性能。

例 6 The design calculations will serve as an illustrative application of the theory of semiconductor devices.

【译文】这些设计计算可用来证明半导体器件理论的实际应用。

例 5 中形容词 familiar 与介词 with 搭配作表语，汉译时则被转译成动词“熟悉”；例 6 中作定语的形容词 illustrative 被转译成相应的动词“证明”。

（3）转译为副词

形容词转译为副词的情况主要有如下三种：

① 有时英语中由形容词所修饰的动作性名词译成了动词，这时该形容词也就相应地需要汉译为副词。

例 7 A continuous increase in the temperature of a gas confined in a container will lead to a continuous increase in the internal pressure within the gas.

【译文】不断提高密闭容器内气体的温度，会使气体的内压力不断增大。

例 8 The modern world is experiencing rapid development of information technique.

【译文】当今世界的信息技术正在迅速地发展。

例 7 中由于两个名词 increase 被汉译为动词“提高”和“增大”，所以相应地需要把它们的定语 continuous 汉译为副词“不断（地）”；例 8 中也因名词 development 被译为动词“发展”，所以作定语的形容词 rapid 也相应译为副词“迅速地”。

② 如果英语中的名词被转译为汉语的动词，那么修饰这个名词的形容词也要转译为相应的副词。

例 9 This graph gives a visual representation of the relationship between these two.

【译文】这份图表可以直观地显示这两者之间的关系。

例 10 With slight modification each type can be used for all three systems.

【译文】每种型号只要稍加改动就能应用于这三个系统。

例 9、例 10 中的名词 representation 和 modification 被汉译为动词“显示”和“改动”，故它们的定语形容词 visual 和 slight 也相应地转译为副词“直观地”和“稍加”。

③ 形容词转译为副词的其他情况。

例 11 The average proportion of molybdenum in igneous rock is about $1.5\times10^{-3}\%$.

【译文】钼在火成岩中所占的比例平均为 $1.5\times10^{-3}\%$ 左右。

例 12 Shortly thereafter, two successful satellites were launched in the United States, Echo I and Telstar I.

【译文】此后不久，又有两颗人造卫星—回声一号和通信一号由美国成功发射升空。

例 13 It is prudent to make sure that there are no short circuits.

【译文】要小心地查明确无短路情况。

4.1.1.3 形容词前增译名词

某些形容词被单独汉译时表示的意义不够明确，需要在前面添加名词，从而以“名词+形容词”这样的主谓短语形式表达出原文形容词的明确含义。

例 14 Unlike satellite communications, or satcoms, standard HF transmitters and receivers can be cheap, light and compact, and require little power to operate.

【译文】与卫星通信系统不同，标准高频发射机和接收机可以做到价格低、重量轻、体积小，并且只需很小的功率即可工作。

例 15 In automotive and architectural design, stresses can be evaluated within the computer without constructing a prototype, making design testing faster and more economical.

【译文】在汽车设计和建筑结构设计中，应力可在计算机中估算，用不着制造样机，从而使设计实验工作进度加快，而且更省钱。

4.1.2 形容词作前置定语的译法

形容词作前置定语即指形容词作定语放在所修饰的名称之前，汉译时一般有以下几种译法：

4.1.2.1 译为“的”字结构

形容词作前置定语时，一般情况下可译为汉语的“的”字结构。

例 16 There are three outstanding scientific changes which will dominate the next fifty years.

【译文】有三个突出的科学变化将会主导未来 50 年。

例 17 Sorting through these sites can be challenging, but there are a number of nationwide job sites you should seriously investigate.

【译文】对所有这些网站进行分类是一项有挑战性的工作，但是有许多全国性的求职网站是应该认真调查的。

4.1.2.2 译为短语

在许多情况下，形容词作前置定语时不能译为“的”字结构，而常与被修饰的名词一起译成短语，或特定的专业技术术语。

例 18 Atoms are much too small to be seen even through the most powerful microscope.

【译文】原子实在太小，即使使用最大放大倍数的显微镜也看不到。

例 19 An incident ray of light is reflected when meeting the surface of a plane mirror.

【译文】入射光线遇平面镜会产生反射。

例 20 Virtual image is an image from which rays of light appear to come. As no light rays actually pass through the image, it cannot be put on a screen.

【译文】虚像是指发出光线的像。而实际上并无光线通过该像，因此该像不能呈现于屏幕上。

例 18 中的 most powerful 不可译为“力量最大的”或“功能最强的”，例 19 中的 incident 若译成“事件，事故”，会令人非常费解，同样，例 20 中的 virtual 若译为“实际上的”，就更让人不知所云。这些形容词应该与它们所修饰的名称一起译为短语或专业术语“最大放大倍数的显微镜”、“入射光线”和“虚像”，才能让读者正确理解。这就需要科技英语翻译工作者在灵活掌握中英文表达方式的同时，还要对翻译文稿所涉及的专业知识有所了解，以免弄巧成拙，贻笑大方。

4.1.2.3 译为主谓结构

有时，形容词作前置定语除了可以译为汉语的“的”字结构以外，还可以根据汉语的语言习惯译为主谓结构。英语句子中表示性状的定语，尤其是由形容词或分词担任的定语，往往具有比较强的谓语性，翻译时很难保留原来的“定语+名词”的形式。此时，如将它们的位置颠倒翻译，即将该定语转译为汉语的谓语（形容词性谓语），构成汉译的主谓短语，则更符合汉语的表达习惯，从而使译文表意更明确，读起来更流畅。

例21 The disadvantage of this method was its extreme slowness since the operation was open looped.

【译文】这种方法的缺点是速度极其缓慢，因为采用的是开环式的操作。

例22 The machine is featured by novel shape, easy operation, high calorific efficiency and low fuel consumption.

【译文】本机的特点是造型新颖、操作简便、热效率高、油耗低。

例23 The reduction in the size of the rotor in turn reduces the size of the surrounding armature resulting in improved mechanical stability and reduced cost.

【译文】转子体积的减小又会减小周围电枢的体积，结果是既提高了机械稳定性又降低了成本。

4.1.3 形容词作后置定语的译法

形容词作后置定语，即作定语的形容词或形容词短语放在所修饰的名词之后的情况，汉译时大多译为汉语的“的”字结构，或视情况灵活处理。

4.1.3.1 “形容词+不定式短语”结构作后置定语的译法

当“形容词+不定式短语”构成形容词短语作后置定语时，可译为汉语的“的”字结构。

例24 Water is a substance suitable for preparation of hydrogen and oxygen.

【译文】水是一种适合于制取氢和氧的物质。

例25 When we do work on something, we have added to it an amount of energy equal to the work done.

【译文】当我们对某物体做功时，我们加入的能量等于所做的功。

4.1.3.2 形容词作不定代词的后置定语的译法

当形容词修饰some、something、any、anything、nothing、everything等词时，形容词作后置定语，常译为汉语的“的”字结构，也可视情况灵活处理。

例26 Now satellite communication is nothing mysterious.

【译文】现在卫星通讯没什么神秘的了。

例27 There is something peculiar that should be noted about the left side of the graph.

【译文】应该注意到曲线图的左边有些特别之处。

4.1.3.3 某些常作后置定语的形容词的译法

以字母 a 开头的形容词和以-able 或-ible 结尾的形容词，还有 enough、present 等词常作后置定语，可译为“的”字结构，也可视情况灵活处理。

例 28 This formula can be found in the physics books available.

【译文】这个公式可以在现有的物理书中找到。

例 29 A laser beam powerful enough and concentrated enough can damage a missile many kilometers away.

【译文】足够强大且足够集中的激光束能够摧毁许多公里之外的导弹。

4.1.3.4 某些形容词并列结构的译法

由 and 或 both . . . and . . . 以及 or 或 either . . . or . . . 连接的两个形容词作定语时常常后置，可以顺译，也可将后置定语译在被修饰名词之前。

例 30 The human body is made up of countless structures both large and small.

【译文】人体是由无数个大大小小的结构组成的。

例 31 Neutron has no charge，neither positive nor negative.

【译文】中子不带电，既不带正电也不带负电。

4.1.4 作表语的形容词的译法

4.1.4.1 直译

形容词在句中作表语时常被直译为汉语的形容词，仍然作表语，或译为形容词谓语。

例 32 Sometimes weight is particularly important.

【译文】有时重量是尤为重要的。或：有时重量特别重要。

如果形容词后带有附加成分，如带介词短语等，这时可译为汉语的“的”字结构。

例 33 Carbon dioxide（CO_2）in the air is mostly responsible for the “greenhouse effect.”

【译文】空气中的二氧化碳（CO_2）是造成“温室效应”的最主要原因。

例 34 Never before in the history of mankind has our world been so rich in scientific discoveries and inventions.

【译文】在人类历史上，我们的世界从来没有过如此富有成果的科学发现和科学发明。

4.1.4.2 转译

有些形容词加 of 的结构可以用相应的动词来替代，转译为动词或名词。

例 35 He is quite ignorant of the machine.

【译文】他对这台机器很不了解。

例 36 Such recent terms as chemical physics and biophysics are indicative of the widening application of the principles of physics，even in the studies of living organisms.

【译文】像化学物理学和生物物理学这类新术语，就是物理学原理被推广应用的明证，表明其应用范围甚至已扩展到生物机理的研究中。

以上两例中，be ignorant of 相当于 not to know，be indicative of 相当于 to indicate，故在译文中译为动词“对……不了解”和名词“明证”。

4.1.5 作状语的形容词的译法

形容词或形容词短语在英语中经常充当原因、让步、结果、时间等状语，翻译时可置于句首，亦可置于句末。

例37　These properties and their effects on the system performance go largely unnoticed.

【译文】这些性质及其对系统性能的影响基本上没有引起人们的注意。

例38　Here k must be a positive number，not equal to zero.

【译文】这里的k必须是一个不为零的正数。

4.2　副词的翻译

副词修饰动词、形容词和其他副词以及整个句子，用来表示时间、地点、程度、原因、条件方式等。依其功能可分为普通副词、疑问副词、关系副词和连接副词。副词在句子中可充当状语、表语和定语。修饰动词时副词的位置比较灵活，可放在句首、句中、句尾；修饰形容词或其他副词时，则要放在所修饰的词之前；作定语时，副词常放在所修饰的名词之后。

普通副词在汉译时大多采用直译法，但有时也需灵活处理。

例39　Usually some of these parameters are known.

【译文】通常这些参数中有一些是已知的。

例40　Computers have been widely used.

【译文】计算机已经得到了广泛的应用。

例41　Conventionally current flowing toward a device is designated as positive.

【译文】习惯上把流向器件的电流指定为正。

例42　This phenomenon is due largely to the skin effect.

【译文】这一现象基本上是由于集肤效应引起的。

例43　The overall thermal efficiency of a nuclear plant is typically 30 percent.

【译文】核电厂总热效率的典型值是30%。

例44　He tried to test the conclusion experimentally.

【译文】他试图用实验方法验证这一结论。

4.2.1　副词作状语的位置

4.2.1.1　副词位于句首

例45　Clearly，many combinations of R_c，R_b，and R_{ie} will satisfy this requirement.

【译文】显然，R_c、R_b、R_{ie}的多种组合能满足这一要求。

例46　Recently astronomers have begun specific research into black holes.

【译文】最近天文学家们开始了对黑洞的具体研究。

4.2.1.2　不及物动词与介词连用时，副词通常位于介词与动词之间

例47　h_{oe} varies exponentially with the collector-emitter voltage.

【译文】h_{oe}随集电极—发射极电压成指数地变化。

例48　At temperatures below the critical temperature，the electrons move freely throughout the lattice.

【译文】当温度低于临界温度时，电子能自由地通过晶格运动。

4.2.1.3 在"形容词+介词"结构中，副词通常位于形容词与介词之间

例 49 This variation of I_{CQ} with temperature is due primarily to variations in V_{BE}.

【译文】I_{CQ}随温度的变化而变化主要是由 V_{BE}引起的。

例 50 We are certain that neutrons act differently from protons.

【译文】我们确信中子的运动不同于质子。

4.2.2 副词作后置定语的几种情况

4.2.2.1 地点副词 above、below、here、there、around、nearby、up、down 等作后置定语

例 51 The two equations above are of great importance.

【译文】上面的两个等式极为重要。

例 52 The table below lists resistivities of some substances.

【译文】下面的表格列出了一些物质的电阻率。

4.2.2.2 时间副词 now、then、today、afterward 等作后置定语

例 53 The difficulty then was the measurement of that parameter.

【译文】当时的困难是如何测量那个参数。

例 54 Scientists today will solve problems tomorrow.

【译文】今天的科学家将解决未来的问题。

4.2.2.3 数量状语+某些副词（away、apart…）

例 55 One end of the rope is tied to a tree 10 ft away.

【译文】绳子的一头被系到10 英尺远的一棵树上。

例 56 The scientists 50 years ago could not do that.

【译文】50 年前的科学家是不可能做到那一点的。

4.2.3 被动语态中副词的位置

4.2.3.1 在没有情态动词的情况下，副词放在过去分词之前

例 57 These parameters are easily measured.

【译文】这些参数容易测得。

4.2.3.2 在有情态动词的情况下，副词可放在 be 之前，也可放在 be 之后

例 58 The reserve voltage gain can usually be neglected.

【译文】反向电压增益通常可以忽略不计。

例 59 Noise may be internally generated.

【译文】噪音可能由内部产生。

4.2.3.3 副词位于被动语态的谓语之后

例 60 The AC and DC components can be treated separately.

【译文】交流和直流分量可以分别加以处理。

4.2.4 副词的其他位置

4.2.4.1 位于主动词之后

例 61　In the following discussion, we usually assume the emf of a source to be constant.

【译文】在下面的讨论中我们通常假设电源电动势是恒定的。

4.2.4.2 位于宾语之后

例 62　We can solve the equations simultaneously.

【译文】我们可以同时解这些方程。

4.2.4.3 位于及物动词之后

例 63　We consider first some special cases.

【译文】我们首先考虑几种特殊情况。

4.2.4.4 在祈使句中可位于动词前或动词后

例 64　Suppose next that the circuit consists of a capacitor.

【译文】下面假设该电路由一个电容器组成。

4.2.4.5 系表结构中一般位于系动词后，若有 will、can、may 等情态动词则位于 be 之前

例 65　h_i is dimensionally an impedance.

【译文】h_i 从量纲上讲是一个阻抗。

例 66　Our logic equation for L will thus be a sum of products containing eight terms.

【译文】因此，L 的逻辑方程是含有八项的乘积之和。

4.2.4.6 修饰不定式时位置较灵活

例 67　It is necessary to decrease propagation delay time dramatically.

【译文】必须大幅度地减少传播延时。

例 68　A better solution is to effectively reduce R_c by replacing it with an active resistance.

【译文】一个更好的解决办法是用一个有源电阻代替 R_c 以有效减少 R_c 数值。

4.3 形容词和副词比较级的翻译

英语中形容词和副词分为原级、比较级和最高级。下面我们就形容词和副词同级比较、优级/劣级比较、最高级比较以及其他一些比较形式的译法加以讨论。

4.3.1 as（not so）+形容词和副词原级+as 的译法

在这一结构中，前面的 as 或 so 是副词，在主句中作状语，修饰其后的形容词或副词原级，后面的 as 是连词，引导比较状语从句。这种结构是形容词或副词的原级比较，表示被比较的双方在性质、程度、数量等方面是相同的。该结构的肯定形式，即“as ... as”句型，一般可译为“和……一样”、“像……那样”；它的否定形式，即“not so ... as”句型，一般可译为“不如……那样”，但翻译时不可生搬硬套，应视上下文灵活翻译。

例 69　The speed of radio waves is as great as that of light.

【译文】无线电波的速度与光速一样快。

例 70 The steam engine was used as widely as the gas engine.

【译文】蒸汽机曾像燃气机那样广泛地使用过。

例 71 The structure of a piston engine is not so simple as that of a turbojet.

【译文】活塞发动机的结构不像涡轮喷气式那么简单。

例 72 The distance of the terminal from the computing center is as great as three kilometers.

【译文】该终端距计算中心的距离远达 3 公里。

原句中 as great as three kilometers 如果译为“像 3 公里一样远”就很不符合汉译的习惯，所以应灵活地译为“远达 3 公里”，读起来就很流畅了。

另外，英语中还有一种表示同级比较的结构，即“no more than”或“no less than”结构，其中的 no 还可用 never 或 nothing 代替，一般被译为“和……一样不……”和“和……一样……”。

例 73 In vacuum light objects fall no slower than heavy objects.

【译文】真空情况下，轻物体与重物体的下落速度一样快。

例 74 The heat of the sun is no less necessary to life than the light.

【译文】太阳的热与太阳光一样，都是生命所必须的。

4.3.2 比较级 + than 的译法

“more . . . than”句型是将两个人或两种事物进行比较时使用最多的句型，句中由 than 引导出一个比较状语从句，或完整或省略某些成分。翻译该句型时不可千篇一律地译为“比…… + 形容词或副词”，应视情况灵活翻译成“与……相比更……”、“……超过了”等，以符合汉语语言习惯。

例 75 The new laboratory costs more than the one built last year.

【译文】新实验室的成本比去年建成的那间高。

例 76 These atoms are separated by distances longer than their diameters.

【译文】这些原子相隔距离的长度超过了它们的直径。

4.3.3 程度状语 + 比较级的译法

有时候，为了表示强调，要在比较级前添加 much、far、even、still、greatly、significantly、considerably、a lot、a little、a great deal 等表示程度的副词和短语，汉译时可译为“更……”、“……得多”、“大大……于”等，或灵活地在译文中体现这些词或短语的含义。如：

例 77 A computer can handle large figures far better than any human brain.

【译文】处理大数字，计算机比任何人脑都好得多。

例 78 Brass expands considerably more than zinc when heated.

【译文】受热时，黄铜膨胀得比锌厉害得多。

4.3.4 more and more 的译法

“more and more”或“ever more”的结构可以表达汉语中“越来越……”的意思。

例 79　The applications of Internet have become more and more popular.

【译文】互联网的应用已经越来越普及了。

例 80　The inflation makes the economy of this region worse and worse.

【译文】通货膨胀使得这一地区的经济越来越糟。

例 81　Nowadays more and more plastic products are used in place of metal.

【译文】现在，越来越多的塑料制品被用来替代金属。

4.3.5　the more ... the more... 的译法

"the more ... the more... "句型一般前面是状语从句，后面是主句，汉译时通常采用顺译法，译为"越……越……"。

例 82　The faster an object moves，the greater is the air resistance.

【译文】物体运动速度越快，空气的阻力越大。

例 83　The finer the diamond particle size，the easier it becomes to plate multiple layers.

【译文】金刚石的粒子越细，就越容易进行多重电镀。

4.4　形容词和副词最高级的翻译

形容词和副词最高级是指在三个或三个以上的人或事物的比较中表示最高、最低、最大、最小等概念。形容词最高级前面通常要加定冠词 the，汉译时将最高级译为"最……"。

例 84　Of all these machines here，this one works most satisfactorily.

【译文】在这里的所有机器中，这台性能最佳。

例 85　Of all the stars in the sky，the sun looks the biggest.

【译文】在天上所有的星星中，太阳看起来最大。

例 86　The most common example of a machine element is a gear.

【译文】机器零件中最普通的例子就是齿轮。

4.5　介词的翻译

英语的介词大多数含义灵活，搭配能力强，一词多义、多用。除了一些具有固定含义的常用介词短语以外，翻译时应遵循汉语的表达习惯，从介词的基本意义出发，联系上下文加以灵活处理。科技英语的介词用法多种多样，其表达习惯与汉语迥然不同。以下介绍几种常用的英语介词翻译方法。

4.5.1　省译法

省译法是在准确表达原文内容的前提下，为了免去译文的累赘表达，省略对英语介词的翻译，以使译文精炼、表意清楚。

4.5.1.1　省译表示时间或地点的介词

有的介词在英语中是不可或缺的，但是译成汉语时则不必译出。比如大部分说明时间或地点的英语介词，译成汉语时如出现在句首往往省略。

例 87　Much progress has been made in electrical engineering in less than a century.

【译文】不到一个世纪，电气工程就取得了很大进展。

例 88　Our country exploded its first atom bomb in 1964。

【译文】1964 年我国爆炸了第一颗原子弹。

例 87、例 88 中说明时间的介词短语汉译时置于句首，均省略了对介词 in 的翻译，分别译为“不到一个世纪”和“1964 年”。

例 89　For a long time scientists could find little use of the material which remained after the oil had been taken out of petroleum.

【译文】科学家们曾长期不能充分利用石油提炼后的残留物质。

原句说明时间的介词词组虽然位于句首，但在翻译时为了符合汉语的表达习惯将其置于主语之后，并省略对介词 for 的汉译。

例 90　As a leaf ages，cells at the base of the petiole react with water.

【译文】树叶衰老时，叶柄底部的细胞与水起反应。

例 91　There might be a massive black hole at the center of our galaxy swallowing up stars at a very rapid rate.

【译文】银河系中心也许有一个巨大的黑洞正以极快的速度吞噬星球。

例 90、例 91 中说明地点的介词短语汉译时均省略了对介词 at 的翻译，分别译为“叶柄底部的”和“银河系中心”，其在译文中的位置按照汉语的表达习惯分别置于句中和句首。

例 92　Researchers in the US have started work on a 100hp（74. 6kW）motor and hope to have it running by the end of next year.

【译文】美国的研究人员已经开始研制一种 100 马力（74. 6 千瓦）的发动机，并希望明年年底让它运行使用。

原句中 on 表示某一具体事物，by 表示时间限制，汉译时可以省译。

4. 5. 1. 2　省译表示与主语有关的介词

例 93　Something has gone wrong with the engine.

【译文】这台发动机出了毛病。

例 94　The chief problem with brush-type motors is the low reliability of the commutator-and-brush assembly.

【译文】电刷型电动机的主要问题是换向器-电刷组合体的可靠性低。

例 93、例 94 中的介词 with 均与主语有关联，汉译时可以省略。例 93 中，with 引导的介词短语汉译时置于句首作主语，译为“这台发动机”；例 94 中，with 引导的介词短语汉译时也置于句首作主语“主要问题”的定语，译为“电刷型电动机的”。

4. 5. 1. 3　省译表示范围、内容的介词

（1）on 表示在某一方面或某一具体事物时可以省译。

例 95　Some authorities are experimenting on the computers based on monitoring systems.

【译文】一些权威人士正在试验监控系统的计算机。

（2）by 表示数值的增减、相差或表示动作执行者且句子又能译为主动句时可以省译。

例 96　The use of hot air instead of cold air reduces the amount of coke needed by more than 70%.

【译文】利用热空气代替冷空气可使所需焦炭量减少 70% 以上。

例 97　Heat and light are given off by the chemical reaction.

【译文】这种化学反应能发出热和光。

原句中介词 by 表示动作执行者，句子可以译成主动句，因而 by 省译。

（3）with 表示伴随状态时可以省译。

例 98　In general, all the metals are good conductors, with silver the best and copper the second.

【译文】一般说来，金属是导体，银最佳，铜次之。

（4）for 表示交换、替代或目的时可以省译。

例 99　The letter T stands for time unless otherwise stated.

【译文】除非另有说明，否则字母 T 就代表时间。

例 100　There is a need for improvement in our work.

【译文】我们的工作需要改进。

（5）in 表示某一方面或处于某种状态时可以省译。

例 101　The temperature in the combustion chamber was in excess of 2,000 °C degrees.

【译文】燃烧室中的温度超过了 2,000 °C。

例 102　Butanes can be maintained in the liquid state at ambient temperature under quite moderate pressure.

【译文】在适当的大气压及常温下，丁烷可保持液体状态。

例 103　Below 4 °C, water is in continuous expansion instead of continuous contraction.

【译文】水在 4 °C 以下不断膨胀，而不是不断收缩。

（6）at 表示速度、温度、成本、比率、价格等时可以省译。

例 104　The workers kept the machine running at a high speed.

【译文】工人让机器高速运转。

（7）over 表示范围时可以省译。

例 105　Nowadays, a typical radio transmitter has a power of 100 kilowatts so that it can broadcast information over a large area of influence.

【译文】如今一台常见的无线电发射机的功率已达 100 千瓦，因而它的播送范围很大。

4.5.1.4　of 的省译

作定语的 of 在汉译时通常译为“……的”，但有时省译 of 更符合汉语的表达习惯。

（1）a lot of、a number of、an amount of、a large quantity of 等说明量的词组中，往往省略对 of 的翻译。

例 106　A large quantity of steam is used by modern industry to generate power.

【译文】现代工业利用大量蒸汽来生产动力。

（2）a kind of、a sort of、a type of 等说明种类的词组中，往往省略对 of 的翻译。

例 107　Nevertheless, this type of cell is still under development.

【译文】无论如何，这种电池还在研制中。

(3) of 短语做定语译成主谓词组或动宾词组时，往往省略对 of 的翻译。

例 108 The change of electrical energy into mechanical energy is done in motors.

【译文】电能变为机械能是通过电动机实现的。

例 109 The nuclear reactions give the sun its constant supply of energy.

【译文】太阳中核反应不断向太阳提供能量。

上述例中的译文省略了对 of 的翻译，分别突出主谓关系的词组“电能变为”和动宾关系的词组“提供能量”。

4.5.2 增译法

英语介词所表达的意义十分丰富，有的介词根据英语规则可以省译，但有时为了使全句意义表达充分，汉译时需要增补一些原文中有其意而无其形的词语。

(1) 翻译介词时增加适当的词，把隐含的意义充分表达出来，使全句符合思维规律，顺理成章。

例 110 The direction is about auto-control system.

【译文】说明是关于自动控制系统方面的。

例 111 Have you heard of the development of a superconductor motor?

【译文】你听说过超导电机研制的情况吗?

上述例中的译文分别增加了“方面”和“情况”两个词，使译文意思明了、易懂。

(2) 英语结构中允许的省略，需要在汉译时适当予以补出，使句子符合汉语习惯，译文清晰、准确。

例 112 The computer can read in and out information in less than a second.

【译文】这台计算机能在不到一秒钟的时间里输入和输出数据。

原句中 and 连接介词 in 和 out，并与动词 read 组成动词短语，在汉译时需要根据汉语的表达习惯在第二个介词 out 前增添“输”字，译成两组并列的词语“输入”和“输出”。

例 113 The oil runs through the pipe, across the pressure pump and into the oiling station.

【译文】油流经管，穿过加压泵，来到加油站。

原句在汉译时增添了“穿”和“来”二字。

(3) 由于汉语表达习惯或修辞的需要，有时要在做定语的介词短词前加译一个动词，使译文更加生动完整。

例 114 They changed several instruments in the system.

【译文】他们更换了安装在此系统中的几个仪器。

例 115 The safety door to the outside was locked.

【译文】通向外面的安全门那时是锁着的。

原句划线部分的介词在汉译时分别增添了动词“安装”、“通向”。

4.5.3 转译法

英汉两种语言在思维表达上存在差异。因此，在忠实于原文的基础上，介词的翻译可以根据汉语的表达习惯转换词性，使汉语译文既通顺又与原文内涵等同。

(1) 英语介词与动词在句法特征上具有共性，即两者都可以带宾语。实际上，不少介词就是由动词演变而来的，并且带有明显的动作意义；还有一些介词虽不直接源于动词，但在一定的上下文中同样具有动词意义。当英语句子中不需要强调动作时常用介词短语结构表达同一概念，这种介词译成汉语时可以译成动词。比如，介词短语表示让步、条件、目的、时间和方式等状语或用做表语、补语时，介词常常转换成汉语的动词。

例 116　For a conventional motor, even with a 95% efficiency, 80kW would be lost as heat.

【译文】一台传统的电动机，即使效率达到 95%，它也将以热能的形式损失 80 千瓦的电能。

原句中的介词 with 与 even 在一起表示让步，译为“即使……达到……”。

例 117　Without friction, there would be no brake.

【译文】没有摩擦，就不会有制动器。

原句中的介词 without 表示条件，译为“没有”。

例 118　Lighting installations have long been a target for energy-saving control schemes.

【译文】照明安装一直是用来控制节能的目标。

原句中的介词 for 表示目的，译成“用来”。

例 119　The machine tool is in good action.

【译文】机床运转良好。

原句中的介词短语作表语，介词 in 译成动词“运转”。

(2) 有些介词译成汉语时常转换成形容词或副词，尤其是一些表示程度和范围的介词，如 about、over、through 等。

例 120　About 1 cm^3 of gas needs to react to produce a current of 1A for one second.

【译文】需要 1 立方厘米左右的煤气参与反应生产每秒 1 安培的电流。

例 121　Like enough, the ship will arrive in port tomorrow.

【译文】这艘船很可能明天进港。

例 120、例 121 中，划线部分的介词 about 和 like 分别转译成了形容词“左右的”和副词“很”。

(3) 介词 with 表示原因、状态时常译成汉语的连词。

例 122　I can hardly do any work with all my reference books taken away.

【译文】我干不了工作了，因为我的参考书全被人拿走了。

例 123　She went away with two lights on in the classroom.

【译文】她走了，但是教室里的电灯却没有关闭。

例 122 中的介词 with 表示原因，汉译成连词“因为”；例 123 中的介词 with 表示伴随状态，汉译成连词“但是”。

4.5.4　反译法

与动词搭配使用的 against、from 等介词短语作表语；besides、but、except 等表示“除了”之意的介词；off、until 等表示时间、范围的介词；above、beyond 等表示品质、行为、能力等“超出……之外”的介词，在翻译时均可以从反面下笔。

例 124 Stay from the bare wire, please.

【译文】请不要靠近裸露的电线。

例 125 Besides coal, we use oil to produce electricity.

【译文】除了煤之外，我们还用油来发电。

例 126 Keep it anywhere but the wet places.

【译文】不要把它放到潮湿的地方。

例 127 The workers lived in a dormitory off the factory.

【译文】工人们住在工厂外面的宿舍里。

例 128 Computers are above acting like a man.

【译文】计算机的行为决不会像人一样。

4.5.5 固定词组的译法

介词翻译切忌望文生义，特别是组成习惯用语时，要多加注意它们的字面意义与实际意义，正确使用其确切含义。

例 129 This big project is subject to many factors such as the man labor and money.

【译文】这个大项目易受到诸如人力和资金等许多因素的影响。

例 130 Without doubt, the assembly operation will focus on low volume products.

【译文】毫无疑问，装配工作将集中在体积小的产品方面。

例 131 In short, like charges repel and unlike charges attract.

【译文】简而言之，相同电荷吸引，不相同电荷排斥。

上述例中，划线部分是介词的固定搭配形式，汉译时要遵循这些习惯用语的实际意义进行。be subject to sth. 译为“易受到……影响”；without doubt 译为“毫无疑问”；in short 译为“简而言之”。

4.6 连词的翻译

连词是连接词与词、短语与短语、句与句的词。英语连词从本身含义和功能上，可分为并列连词和从属连词两类，从形式上又可分为简单连词（and、but、for、before...）、关联连词（both... and...、not only... but also...）、分词连词（provided、supposing...）和短语连词（as soon as, so that...）四类。

4.6.1 连词的一般译法

考虑到连词本身的意义和汉语的表达习惯，连词的译法一般有如下三种：

4.6.1.1 省译

英语中，无论是时间状语从句、原因状语从句或条件状语从句，还是表语从句或宾语从句等名词性从句，均需要相应的从属连词来引导。汉译时，一方面，一些英语连词在句中只起连接作用而本身并无意义；另一方面，汉语常依靠词序把两个或更多句子成分连起来，而不使用连词。因此，这样的连词常可酌情省译。

（1）省译并列连词

例 132 Electric charges, positive and negative, which are responsible for electrical force, can wipe one another out and disappear.

【译文】产生电场力的正负电荷会相互抵消掉。

例 133 Bacteria, even great in number, are invisible to the unaided eye, but they can easily be distinguished by the microscope.

【译文】细菌即使为数众多，肉眼也是看不见的，借助于显微镜却容易辨认出来。

例 132、例 133 中分别省译了连词 and 和 but。

（2）省译从属连词

例 134 As the temperature increases, the volume of water becomes greater.

【译文】温度升高，水的体积就增大。

例 135 Because energy can be changed from one form into another, electricity can be changed into heat energy, mechanical energy, light energy, etc.

【译文】能量能从一种形式转换为另一种形式，所以电可以转变为热能、机械能、光能等。

例 136 Providing that we know the current and resistance, the voltage can be calculated.

【译文】知道了电流和电阻，就能求出电压。

例 137 The truth is that the current increases with every decrease of resistance.

【译文】事实上，电流随着电阻减小而增大。

例 136、例 137 中分别省译了从属连词 as、because、providing that 和 that。

4.6.1.2 直译

有些连词本身具有一定含义，如果省译会影响对句子的理解，这时就要采取直译法，把连词的字面意思直接翻译出来。

例 138 The wire must not be in contact with the air or it will become oxidized and burn out.

【译文】金属丝不应与空气相接触，否则会被氧化并烧毁。

例 139 Both teaching and research work are making great strides.

【译文】教学与科研都在大踏步前进。

4.6.1.3 转译

为符合汉语表达习惯并明确传递原文含义，汉译时英语的连词也可进行转译。这里所谓的转译具有两重含义：一是连词间的转译，即并列连词有时可以转译为从属连词，而从属连词有时也可转译为并列连词，或从属连词改变原来的功能转译为其他意义的从属连词；二是连词与其他连词间的转译，即根据连词在句中的关联作用有时可以转译为汉语中的副词、介词、助词以及动词。

例 140 Every object, large or small, has a tendency to move toward every other object.

【译文】每一个物体，不论大小，都有向其他物体移动的倾向。

原句中的并列连词 or 转译为“不论”，表示让步。

例 141 The atmosphere, therefore, is heated by contact with the surface and vertical motion thereby ensures.

【译文】这样，大气由于与地面接触而被加热，从而引起了垂直对流。

例 142 This material is ductile and malleable.

【译文】这种材料既有韧性又有延展性。

例 141、例 142 中的并列连词 and 分别转译为汉语中表结果的连词“从而”和副词“又”。

例 143 Cracks will come out clean when treated by ultrasonic waves.

【译文】如果以超声波处理，缝隙就会变得很洁净。

原句中的从属连词 when 转译为“如果”，表示条件。

例 144 The materials are excellent for use where the value of the work pieces is not high.

【译文】如果零件价值不高，最好使用这种材料。

原句中的从属连词 where 转译为“如果”，表示条件。

4.6.2 常用连词的译法

4.6.2.1 并列连词的译法

本节选择科技英语中最为常见而翻译时又较难处理的五种并列连词：and、or、but、as well as 和 either... or...，说明其具体译法。

（1） and 的译法

① and 在句中连接两个对等的并列成分时，可译为“和”、“与”、“并”、“及”等。有时也可省译。

例 145 Computers cope with negative numbers and fractions in binary arithmetic.

【译文】计算机采用二进制算法来处理负数与小数。

例 146 Some electrical engineers design and maintain power plants, transmission lines and home and factory electrical installations.

【译文】有些电气工程师设计和保养发电厂、输电线、以及各种家用和工业用电器设备。

例 147 Oxygen is chemically active and takes part in many reactions.

【译文】氧的化学性能活泼，能参与许多反应。

原句中省略了对 and 的翻译，而把原句译成两个对等的分句。

② and 连接的句子成分有对比意义时，可译为“而”或省译。

例 148 The induced electromotive force lags and the current leads.

【译文】感应电动势滞后，而电流导前。

例 149 Like charges repel and unlike charges attract.

【译文】同性电荷相斥，异性电荷相吸。

译文中的“同性电荷”与“异性电荷”是一组具有对比意义的词组，因此省译 and。

③ and 或 and yet 表示转折时，可译为“但”、“可是”、“然而”。

例 150 Transformers are commonly used in radio receivers and electrical communication circuits, and they are generally much smaller than those for power transmission.

【译文】变压器通常用于无线电收音机和电信电路中，可是这些变压器一般比用于输电的变压器小很多。

例 151 Most colds are caused by viruses, and yet antibiotics can't harm viruses at all.

【译文】大多数感冒是由病毒引起的，然而抗生素对杀灭病毒丝毫不起作用。

④ and 表示递进关系时，可译为“而且”、“并且”、“再”；表示并行关系时，可译为“既……又……”、“又……又……”。

例 152　It is important to know that different types of control circuit may be used in one control system, and a completed control loop may consist of a combination of electric, pneumatic, or hydraulic elements.

【译文】重要的是要知道，在一个控制系统中可使用不同类型的控制电路，而且一个完整的控制回路可由电力、气动或液压配合组成。

例 153　The projection is irregular and variable.

【译文】突出部分既不规则又易变化。

⑤ and 连接两个动词时，可酌情译为“来”、“以（便）”或“就”等。

例 154　The plant requires nitrogen and makes proteins.

【译文】植物需要氮气来制造蛋白质。

原句中的连词 and 后面的行为 make proteins 是 and 前面的行为 require nitrogen 的目的，因此译为“来”更符合原句所表达的意思。

例 155　Change one or more steps and we will improve the quality of the finished products.

【译文】若改变一个或几个步骤，我们就有可能提高成品的质量。

原句中的连词 and 前面的行为 change one or more steps 是 and 后面的行为 will improve the quality of the finished products 的条件，因此译为“就”更符合原句所表达的意思。

⑥ and 连接两个含因果关系的并列句时，可译为“因此”、“因而”、“从而”等。

例 156　Lead lacks tensile strength and it cannot be drawn out in the form of fine wire.

【译文】铅的张力强，因此不能制成细丝状。

例 157　For LP turbine there is no residual axial thrust caused from blading because of the turbine's double-flow arrangement of blading and thus no balancing piston is needed.

【译文】至于低压汽轮机，并没有残存的轴力，因为汽轮机的叶栅作双流配置，因而也无需采用平衡轮。

（2）or 的译法

① or 连接两个对等的词或短语，表示选择其一，可译为“或”、“或者”。

例 158　No electric charges have been observed of smaller magnitude than charge of protons or electrons.

【译文】还没有观察到比质子或电子的电量更小的电荷。

② or 连接两个同义词语或同一事物的两种不同的表述，可译为“即”、“也就是”、“等于”。

例 159　Scientists have shown that certain chemicals are slowly eroding the ozone layer. The main culprits are chlorofluorocarbons or CFCs.

【译文】科学家们已经证明，某些化工产品会慢慢破坏臭氧层，主要诱因就是氯氟甲烷，即氟里昂。

例 160　Fire is a chemical action, or an action taking place between different chemical substances.

【译文】火是一种化学反应，也就是发生在不同化学物质间的一种作用。

③ or 连接的短语表示让步概念时，可译为“无论”。

例 161 Right or wrong，his theory is the most influential.

【译文】无论对错，他的理论无疑是最有影响力的。

④ or 或 or else 连接句子时，可译为“否则”。

例 162 A body must move，or no work is done.

【译文】物体必须移动，否则就没有做功。

例 163 You have to open the switch right away or else the fuse will be burned off.

【译文】你必须把开关立即关掉，否则保险丝就会烧断。

⑤ or 连接数字时表示不确定，可译为“大约”、“左右”。

例 164 The particles require one or two days to reach the earth.

【译文】这些粒子到达地球大约需要一两天。

例 165 The newly-installed generating set will be put in operation a month or so.

【译文】新机组再有一个月左右就投入运行了。

（3）but 的译法

① but 单独使用时通常表示转折关系，可译为“但”、“可是”、“然而”等。但是 not... but... 表示并列关系，可译为“不是……而是……”。

例 166 The plan is very complicated technically，but the general idea is quite simple.

【译文】这个计划在技术上十分复杂，但总的想法却相当简单。

例 167 Bionics is not a specialized science but an inter-science discipline.

【译文】仿生学不是一门单独的学科，而是一门边缘学科。

② but that + 从句，后接虚拟句型，可译为“假使……不”、“要不是”等。

例 168 But that you cut the engine immediately，it would have burned out.

【译文】假使你不及时关掉发动机，那么它早就坏了。

例 169 But that he prevented me，I might have kept the machine parts in the damp place.

【译文】若不是他阻止，我可能就会把那些机器零件保存在潮湿的地方。

③ but for + 名词（代词），用于虚构的条件句，可译为“如果没有”等。

例 170 But for the laws of motion as a basis，thousands of discoveries and inventions would have been out of the question.

【译文】如果没有运动定律作为基础，就不可能有成千上万的发现和发明。

（4）as well as 的译法

当 as well as 置于句首或句中列举同类句子成分时，可译为“以及”；置于句末时，要强调 as well as 前面的词语，可以使用倒译法，译为“不仅……而且……”。使用顺译法时，可译为“既……也……”。

例 171 Boiling point elevations，as well as the variation of the heat capacities and latent heats with temperature，are negligible.

【译文】沸点升高以及热容和潜热随温度的变化忽略不计。

例 172 The physics students study chemistry as well as physics.

【译文】物理系学生不仅学物理，而且学化学。

例 173 In order to get various vitamins a person should eat a diet including vegetables as well as meat.

【译文】为了摄取多种维生素，一个人既应该吃些素食，也该吃荤食。

例172、例173中的as well as置于句末，分别采用倒译法，译为“不仅……而且……”和顺译法，译为“既……也……”。

（5）either... or 的译法

or与either连用主要是为前后呼应加强语气，可译为“（或）……或……”、“是……还是……”等。

例174 When light strikes an opaque object, the light is either absorbed or reflected.

【译文】当光照在不透明的物体上时，光或是被吸收或是被反射。

例175 When astronomers look at the sky, they expect to see either a star or a galaxy.

【译文】每次天文学家观测天空，他们期望看到的不外乎是一个恒星或一个星系。

对应地，并列连词neither... nor... 一般译为“既非……也非……”或“既不……也不……”。

4.6.2.2 从属连词的译法

在科技英语中使用从属连词的频率非常高，特别是when、before、as、while、until（till）和as soon as等。

（1）when 的译法

例176 When a substance is dissolved in water, the freezing point of the water is lowered.

【译文】当一种物质溶解于水时，水的冰点就会降低。

例177 Fire at the enemy when you see my signal.

【译文】一看到信号就向敌人开火。

例176、例177中when引导时间状语从句，分别译为“当……时”和“一……就……”，例177的译法更强调时间上的紧凑。

例178 An electric current exists when there are charges in motion.

【译文】如果有电荷运动，就存在着电流。

原句中when引导的从句除表示时间外，按上下文的逻辑意义还表示条件，因此when酌情译为“如果……就”。

例179 We don't know when the best times are for seeing Mercury.

【译文】我们不知道观察水星的最好时机是什么时候。

when引导名词性从句时仍保留疑问含义，原句中when译为“是什么时候”，其他情况还可译为“在什么时候”、“何时”。

例180 In those years when electricity had not been discovered, people knew nothing of conductors and insulators.

【译文】在电还没有被发现的年代里，人们不知道有导体和绝缘体。

when可以转化为关系副词，引导某个时间（段）的定语从句。原句中when译为“……的”。

（2）before 的译法

例181 Before you join one electric wire with another, make sure the current has been switched off.

【译文】在你把一根电线和另一根电线连接起来之前，务必先切断电源。

原句中 before 引导时间状语从句译为“在……之前”。

例 182 Lubricate each movable part before it wears away.

【译文】要对每个运动部件进行润滑，以免磨损。

原句中将 before 译为“以免”，强调反面后果，这是对 before 引导时间状语从句的灵活处理。

例 183 Further study is needed before the origin of the covalent bond can be considered a settle question.

【译文】只有进一步深入研究，才能解决共价键的起源问题。

当主句与从句中含有 must、can 等情态动词时，before 译为“只有……才能……”，更加强调主、从句动作的先后次序。

例 184 She switched on the engine before we could stop her.

【译文】我们还未来得及阻止她，她就开动了发动机。

原句在汉译时使用倒译法，将 before 译为“未……就……”，强调从句动作晚于主句动作。

例 185 The direction of the current was decided before electrons were discovered.

【译文】电流方向确定了，电子才被发现。

原句在汉译时使用顺译法，同样强调从句动作晚于主句动作，将 before 译为“（后）……才……”。

（3） as 的译法

例 186 As current is increased, the flux builds up according to Curve 2.

【译文】当电流增大时，磁通沿着曲线 2 增大。

原句中 as 引导时间状语从句，译为“当……时”，也可译为“随着”，表示动作正在进行，即：随着电流增大，磁通沿着曲线 2 增大。

例 187 As you are doing experiment, write down the observations.

【译文】一面做实验一面将观察到的情况记下来。

原句中 as 译为“一面（边）……一面（边）……”，表示动作同时发生。

例 188 As the current was trebled abruptly, the second valve burnt out.

【译文】由于电流突然增加了两倍，第二个电子管烧坏了。

原句中 as 引导原因状语从句，译为“由于（因为）”。

例 189 Gravity, the force of the earth's attraction, governs the atmosphere just as it governs solid objects.

【译文】引力，即地球的吸引力，正像它影响着固体那样影响着大气。

原句中 as 引导方式状语从句，并与 just 连用，译为“正像……那样”，也常译为“正如……一样”。

例 190 Electron tubes are not so light in weight as semiconductor devices.

【译文】电子管的重量不如半导体器件那么轻。

原句中 as 引导比较状语从句，“not so... as...”结构可译为“不如……那么……”或“没有……那样”。对应地，“as... as...”结构表示肯定的比较，可译为“和……一样”。

例 191　The loop oscillator frequency can be the same as, or a multiple, the reference frequency.

【译文】环路振荡器可以和参考频率相同，也可以是参考频率的倍数。

原句中的 the same as 译为“和……相同”。类似结构还有“such... as...”，译为“像……一样（的）”。

（4）while 的译法

① while 作并列连词的译法

例 192　Motion is absolute while stagnation is relative.

【译文】运动是绝对的，而静止是相对的。

原句中 while 连接两个并列句，表示对比，相当于 while as，译为“而”。

例 193　New types of computers are smaller in size, while they are simpler in operation.

【译文】新型计算机体积较小，而且操作也比较简单。

原句中 while 作为并列连词，表示递进，相当于 and what is more，译为“而且”。

② while 作从属连词的译法

例 194　While the moon is revolving around the earth, the earth is also revolving around the sun.

【译文】在月球绕着地球旋转的同时，地球也在绕着太阳旋转。

例 195　While there are heat losses in a steam turbine, I cannot prove it.

【译文】虽然涡轮机内有热损耗，但我无法加以证明。

原句中的 while 译为“虽然”，也可译为“尽管”，表示让步。

例 196　I'd like to get it settled today while we are at it.

【译文】既然我们着手干了，我想今天就把它干完。

原句中的 while 译为“既然”，相当于 since，表示原因。

（5）until（till）的译法

例 197　When the water temperature is increased, it vaporizes more quickly until it reach boiling point.

【译文】水温升高时，水蒸发得越来越快，直到达到沸点为止。

例 198　Not until the jet engine had come into use could planes travel at supersonic speeds.

【译文】直到喷气发动机投入使用，飞机才能做超音速飞行。

until（till）引导时间状语从句，有两种译法。例 197 的主句为肯定句，until 译为“直到……为止”；例 198 的主句为否定句，until 译为“直到……（后）……才……”，有时也可译为“在……以前……不”。

（6）as soon as 的译法

例 199　As soon as the electrons are transformed and ions formed, they are charged.

【译文】一旦电子转移，形成离子，它们就带电了。

例 200　The electrons create a magnetic field as soon as they begin to move.

【译文】电子一开始运动，就产生磁场。

例 199、例 200 中，as soon as 均引导时间状语从句，分别译为“一旦……就……”和“一……就……”。

4.7 冠词的翻译

冠词是英语特有的一种词类。英语冠词数量虽少，使用频率却很高，其用法也非常纷繁复杂。当冠词在英语句子中不表达具体的词义，而只是为了满足语法对用词造句形式上的要求时，这种冠词经常省略不译。然而，当冠词在英语句子中不但起到语法作用，而且表达某种实际意义时，这种冠词一般不宜省译。

4.7.1 冠词的一般译法

4.7.1.1 定冠词的一般译法

当定冠词具有词汇意义，起着指示代词（this、that、these）的作用时，通常要在译文中体现其词汇意义。如定冠词 the 表示某一类特定的人或事物中的“某一个”时，可译为“该”、“这个/种”、“那个/种”等，当定冠词 the 后面的名词是复数时，则译为“这些”、“那些”。

例 201 The machines over there have been well lubricated.

【译文】那边那些机器都已很好地加过润滑油了。

例 202 The alloy can be electroplated or otherwise processed.

【译文】该合金可以电镀，或者用其他方式处理。

4.7.1.2 不定冠词的一般译法

不定冠词 a 和 an 修饰单数名词。不定冠词的可译性比定冠词强，尤其是在不定冠词明显地表示数量“一”，或是表示“每一”、“同一”时，往往不宜将其省译。

例 203 The electric resistance is measured in ohms. An ohm is a volt per ampere.

【译文】电阻的测量单位是欧姆。一欧姆是每安培一伏特。

例 204 The number of vibrations a second is called frequency.

【译文】每秒振动次数称为频率。

例 205 The solution should be changed a month.

【译文】每月应更换溶液一次。

4.7.2 冠词的省译

4.7.2.1 定冠词的省译

英语定冠词在泛指类别，表示世界上独一无二的事物，或者用于带有限定性定语的名词之前、形容词最高级或序数词等词语之前时，一般省略不译。

例 206 The proton has a positive charge and the electron a negative charge, but the neutron has neither.

【译文】质子带正电，电子带负电，中子两种电荷都不带。

例 207 The pump is one of the oldest machines.

【译文】泵是最古老的机器之一。

例 208 As we know, the sun is one of the greatest sources of electromagnetic waves.

【译文】正如我们所知，太阳是最大的电磁波源之一。

例 209　The color sensation produced by light depends simply upon the length of the wave producing the light.

【译文】光所产生的颜色，只是由产生光的波长所决定。

4.7.2.2　不定冠词的省译

（1）不定冠词在泛指某一类事物中的任何一个时，往往省略不译。

例 210　A transformer must have two coils.

【译文】变压器一定有两个线圈。

例 211　A material whose deformation vanishes rapidly with the disappearance of the loads is said to behave elastically.

【译文】随着荷载的消失而变形也很快消失的材料，据说是具有弹性。

例 212　Ohm's law indicates that whenever a current flows through a resistance, a difference of potential exists at the two ends of that resistance.

【译文】欧姆定律指出：只要有电流流过电阻，该电阻的两端就有电位差。

（2）英语中有很多含有不定冠词的常用短语。汉译时，这种短语中的不定冠词一般省译。如：

a large amount (number) of　大量；　as a matter of fact　事实上，其实
for a time　一些时候，暂时；　for an instant　一瞬间，片刻
in a sense　在某种意义上；　make an exception of　将……除外
on a fifty-fifty basis　平均地，在对开分的基础上；　without a hitch　无障碍地
to a certain extent　有几分；　with a view to　为了，意在

4.8　英语倍数、分数、百分数增减及比较的翻译

数字和量词的使用在科技英语中是一个重要的构成部分。在当今信息时代里，获取信息不仅要快，而且要准确。如果不能正确地理解原文的意思，翻译的数据不准确，就会带来不可估量的损失。

英汉两种语言在倍数、分数和百分数增减的表达方式存在一定的差异。英语倍数增减有多种表达法，但都包括基数；汉语在这方面也有不同的表达方法，有的句式（或措词）包括基数，有的不包括基数。因此，在进行英语倍数的汉译时，必须根据两种语言的特点准确无误地显示出包括或不包括基数，必须与原文所指的实际倍数相吻合。

4.8.1　具体数字增减及比较的译法

人们比较容易接受的是用具体数字表示数量增减，如用 increase、rise、go up、drop、decrease 和 fall 等代表有增减意义的动词与"by + 纯增减量"表示"增加或减少了……"。其他几种形式有"increase from N_1 to N_2"（从 N_1 增加到 N_2），"系动词 + N + more than"，等等，与汉语习惯无甚差别，因此，均可采用直译的方法。

例 213　Hot gas is fed to the colder, where its temperature drops to 20 °C.

【译文】将热气体送到冷却室，其温度降到 20 °C。

例 214　The number of the countries having diplomatic relations with China has increased to more than 100.

【译文】和中国建立外交关系的国家已增至100多个。

例215 When the sun goes down, the temperature on the moon may be as low as 160 °C below zero.

【译文】太阳落下之后，月球上的温度可低达零下160摄氏度。

"as + 形容词 + as + 数字"这一句型一般可译为"……达 + 数字"。

例216 Brazil is larger than the Continental United States by about 185,000 square miles.

【译文】巴西比美国大陆大18.5万平方英里左右。

4.8.2 倍数增减及比较的译法

4.8.2.1 倍数增加或比较的译法

（1）包括基数的增加的表达和译法

英语中说"增加了多少倍"，一般采用以下表达方式：

①"N times + as + 形容词原级 + as"和"N times + 形容词比较级 + than"

上述数量范围都是连基数也包括在内的，是表示增减后的结果；而在汉语里所谓"增加了多少倍"，则只表示纯粹增加的数量。因此，对英语中表示倍数增加的各种方式，翻译时都可用以下两种说法表达："增加了 $N-1$ 倍"（表示净增，不包括基数）；"增加到 N 倍"或"是原来的 N 倍"（包括基数）。

例217 The production of various stereo recorders has been increased four times as against 1977.

【译文】各种立体声录音机的产量比1977年增加了三倍。

例218 The output of color television receivers increased by a factor of 3 last year.

【译文】去年彩色电视接收机的产量增加了二倍。

例219 In case of electronic scanning the bandwidth is broader by a factor of two.

【译文】如用电子扫描时，带宽增加一倍。

例220 The speed exceeds the average speed by a factor of 3.5.

【译文】该速度超过平均速度的两倍半。

上述3个例子中，N 均大于1。若 N 为带分数或小数的情况，则处理为百分比更符合汉语习惯。如 increase by a factor of 1.25 可译作"增加了25%"。但 N 本身为一个大于零小于1的百分数或纯分数，如 increase by a factor of 1/5，increase by three to six percent，数字又为净增数值时，分别译为"增加20%"和"增加3%～6%"。

例221 This machine is five times heavier than that one.

【译文】这台机器比那台机器重四倍。

例222 This railway is three times as long as that one.

【译文】这条铁路的长度是那条铁路的三倍。

例223 The melting point of titanium is almost twice as high as that of platinum.

【译文】钛的熔点差不多是铂的熔点的两倍高。

例224 Iron is almost three times as heavy as aluminum.

【译文】铁的重量几乎是铝的三倍。

英语在倍数比较的表达上，其传统习惯是 more than 等于 as much as，汉译时不能只从字面上理解，将其译为“比……大 *N* 倍”，而应将其译为“是……的 *N* 倍”，或“比……大 *N* - 1 倍”。“*N* times + as + 原级 + as + 被比较对象”，表示“是……的 *N* 倍”。

② 表示成倍增加的动词

例 225 Within the next forty years, the world population may double.

【译文】在今后的 40 年内，世界人口可能增加一倍。

例 226 If you treble the distance between an object and the earth, the gravitational attraction gets nine times weaker.

【译文】如果把一个物体与地球的距离增加两倍，地心引力就会减弱到九分之一。

例 227 If the resistance is doubled without changing the voltage, the current becomes only half as strong.

【译文】若电压不变，电阻增加一倍，电流就减小一半。

例 228 As the high voltage was abruptly trebled all the valves burnt.

【译文】由于高压突然增加了两倍，所有的电子管都烧坏了。

double 和 twice 表示倍数，应译为“增加一倍”或“是……的两倍”。类似的表达方式还有用动词 quadruple，duplicate 等，分别表示“是……的四倍”、“是……的两倍”或“比……增加了三倍”、“比……多一倍”。

例 229 The number of farms in the United States tripled between 1860 and 1910.

【译文】1910 年，美国农场的数目是 1860 年的三倍。

③ be + 形容词比较级 + by + a fact of + *N* + . . .

例 230 In the case of electronic scanning the beam width is broader by a factor of two.

【译文】在电子扫描情况下，光带宽度将增大一倍。

④ be + *N*-fold + more than + . . .

例 231 Under high pressure the particles beneath the crust are 2000-fold more than normal.

【译文】高压下表壳下的粒子数是常压下的 2000 倍。

例 232 New booster can increase the payload by ten-fold.

【译文】新型助推器能使有效负载增加 10 倍。

⑤ 表示增长或提高的动词 + 倍数 + ……

例 233 The production of integrated circuits has risen to 4 times as compared with last year.

【译文】集成电路的产量比去年增加了三倍。

例 234 The sales of industrial electronic products have multiplied 6 times since 1999.

【译文】自 1999 年以来，工业电工产品的销售量增加了五倍。

例 235 The speed exceeds the average speed by a factor of 2. 5.

【译文】该速度超过平均速度1. 5 倍。

例 236 The error probability of binary AM exceeds binary FM at least 6 times.

【译文】二进制调幅的误差比二进制调频至少大 5 倍。

例 233 ~ 236 中的表达方式是包含基数在内的，不是指净增数，而是指增加后的结果。按照汉语意思，都有“超过”或“提高”的含义，根据汉语的表达习惯，是不包括基数在内的，所以在译成汉语时一定要在原文的倍数上减 1，即译成 *N* - 1 倍。

此外，英语中表达成倍增加的常用词组还有 go up half over...（比……增加一半）、increase by N powers of ten（增加 10N 倍）、N times as much as... the same（是……的 N 倍）等。

（2）不包括基数增加或比较的表达和译法

在英语关于倍数增加的表示中，不包括基数增加或比较的表达形式只有一种：... + as + adj. + again + as...。

例 237 The leads of the new condenser are as long again as those of the old.

【译文】新型电容器的引线长度比旧式的加长了一倍。

例 238 The output of this one is half as powerful again as that one.

【译文】这台（电机）的输出力比那台大一半。

例 239 Wheel A turns as fast again as wheel B.

【译文】A 轮转动比 B 轮快一倍。

例 240 The input is as great again as the output.

【译文】输入功率比输出功率大一倍。

以上各例在翻译时，都将原句的数量关系结构译为净增长值。

4.8.2.2 倍数减少、比较的译法

英语表示减少的基本句式与表示增长的倍数的基本句式类似。英语中可以用倍数来表示减少的程度，如 decrease/reduce/drop + by + N times，less than + N times。翻译时，因为汉语不说“减少了几倍”，所以通常要把倍数换算成分数或百分数来表示：表示净减数时为“减少了（N－1）/N”；包括基数成分（表示剩下）时为“减少到 1/N”。

例 241 The automatic assembly line can shorten the assembling period (by) ten times.

【译文】自动装配线能够将装配期缩短十分之九。

例 242 This metal is three times as light as that one.

【译文】这种金属比那种金属轻三分之二。

例 243 The dosage for a child is sometimes twice less than that for an adult.

【译文】小孩的剂量有时为成年人剂量的三分之一。

或小孩的剂量有时比成年人的剂量少三分之二。

例 244 Since the introduction of new technique the switching time of the transistor has been shortened three times.

【译文】采用这项新工艺后，晶体管的开关时间缩短了三分之二。

例 245 When the signal has increased by 10 times the gain may have been reduced by 8 times.

【译文】如果信号增大九倍，增益就可能降低八分之七。

例 246 The moon is eighty times lighter than the earth.

【译文】月球的重量只有地球的八十分之一。

例 247 The new motor is five times as light as the old one.

【译文】新电机比旧电机轻五分之四。

上述句子在翻译时，都按照汉语的表达习惯是把减少后的数量用分数来表示，译成“减少了（N－1）/N”和“减少到 1/N”或“是（原来）的 1/N”。

4.8.3 百分数增减及比较的译法

英语中百分数增减及比较的表达常采用“表示增减的动词 + to + 百分数”和“百分数(+ of) + 名词(或代词)”两种句型表示增减后的结果，包括底数在内。翻译时可以直接译出，或按倍数译法处理。

例 248 Using the new process, the loss of metal can be reduced to 20 per cent.

【译文】如果采用这种新工艺，可使金属损耗下降到20%。

例 249 In 1933, industrial output in the USA fell to 65 per cent of the output in 1929.

【译文】1933年，美国的工业产值降至1929年的65%。

上述例中的百分数表示增减后的结果，包括底数在内，翻译时直接译出。

例 250 The boiler cuts its coal intake by 22 per cent when scales were removed.

【译文】水垢清除后，锅炉的耗煤量减少了22%。

例 251 The United States Census Bureau notes that the population of the world still is expected to increase by more than 25% in the next 13 years.

【译文】美国人口普查局指出，在今后13年中，世界人口预计将增加25%。

由“by + 百分数和分数”，无论增减，一般均表示净增减部分，可以照译。

4.8.4 不确切倍数的译法

如果倍数是一个很大的近似值，相差一倍并无多大影响时，翻译时则无须减去1。

例 252 Synchrotron radiation in the X-ray range would be 1 million times brighter than that from an X-ray tube.

【译文】同步加速器辐射的亮度比X射线管所发出的大100万倍。

例 253 The sun is 330,000 times more massive and a million times more bulky than the earth.

【译文】太阳的质量比地球大330,000倍，体积比地球大100万倍。

例 254 The thermal conductivity of metals is as much as several hundred times that of glass.

【译文】金属的导热率比玻璃高数百倍。

总之，英汉两种语言在数量表达上的差异是比较复杂的。这里涉及语言的表达习惯和理解的问题，这一点需在翻译实践中不断吸取经验。但是无论如何，准确无误地翻译出原文的意思才是翻译的首要任务。

翻译练习

1. Complex computer systems are prone to software hitches.
2. Ice floats because it is not as dense as water.
3. This in a large measure is responsible for the high accuracy obtainable from such systems.
4. Research and development is of particular importance to aerospace industry.
5. Lubricating oils are used to ensure the smooth running of the machine.
6. The World Wide Web is a unique medium for communication and for publishing.
7. A practical means of transferring geothermal energy is to convert it to electric power.

8. Needle bearings have a high load capacity.

9. The gear teeth should have the strength sufficient to operate.

10. Scientists have found a method of measurement suitable for most island water and coastal areas.

11. The resistance of the load is too high for the amplifier.

12. Deserted soils are sparsely covered with shrubs and grass.

13. Normally the whole mass of the atom is concentrated in the nucleus.

14. The most important advantages of rolling bearings over sliding bearings is the most complete elimination of friction.

15. To detect gases, a sample of air can be analyzed chemically.

16. The salts should be heated electrically, if possible.

17. Metals do not change their forms as easily as plastic bodies do.

18. This material can withstand higher stresses under repeated loading.

19. The greater the cross-sectional area, the greater the strength will be.

20. Hydrogen is the lightest gas, its density being 0.0894 g/l.

21. Many countries are investing nothing at the moment.

22. These machines have been used extensively and very successfully for some years especially in the USA.

23. At the surface of a liquid there is a surface tension.

24. Gold is similar in color to brass.

25. There are many substances through which electric currents will not flow at all.

26. The expense of this project is not a small sum.

27. The fence by the cable ditch is too low.

28. Metals with which people make machines were discovered long ago.

29. The high temperature keeps us out of the room.

30. Throughout the world come into use the same signs and symbols of mathematics.

31. I can't concentrate on my work with so many machines making so much noise.

32. The ship sails well except that it was a little loaded.

33. This machine has worked in succession for seven or eight hours.

34. This shows that something unexpected may have turned up.

35. Where the watt is too small a unit, we may use the kilowatt.

36. Sometimes it is necessary to drill to a depth of 1,000 meters or more, before water is found in sufficient quantity.

37. When short waves are sent out and meet an obstacle they are reflected.

38. Because this current continually reverses in direction, it is called an alternating current.

39. Some companies replace certain parts regularly, determining the effective life of a part and then replacing it just before it would wear out.

40. That is why surfactants are so widely used in industry.

41. Oxygen is chemically active and takes part in many reactions.

42. Modern electronic computers and other devices will be made smaller by the development of LSI.

43. The words "suitably selected" should be duly stressed.

44. The machine is intended for grinding the top and bottom surfaces of tungsten carbide tips.

45. A polymer is a substance of high molecular weight.

46. These machines may be oiled once or twice a month.

47. The rate of a chemical reaction is proportional to the concentrations of the reacting substances.

48. A light year is the greatest unit of measurement for distance.
49. Any substance is made up of atoms whether it is a solid, a liquid, or a gas.
50. The voltage has dropped by five times.
51. C is twice less than D.
52. The principal advantage over the old fashioned machine is a three-fold reduction in size.
53. The capital in I has grown from 6,000 to 6,500, or by 1/12.
54. This kind of film is 3 times thinner than ordinary paper.
55. The dose of dexamethasone is seven times smaller than that of prednisone.
56. The prices of medicine have been reduced 4 times as compared with 1950.
57. A yard is three times larger than a foot.
58. The new compressor is half as heavy as the old one.
59. Reducing the data rate by one-half will double the duration of each symbol interval.

第5章　短语的翻译

科技文章要求行文简练、结构紧凑、句式严整。因此，经常使用分词短语、介词短语和不定式短语等结构代替从句，这些非谓语动词及短语使句子言简意赅、主次分明、结构严谨、逻辑性强。

5.1　分词短语的翻译

分词是英语动词非限定形式中的一种。分词短语分为现在分词短语和过去分词短语，它们在句子中担任的成分大体上是相同的，其主要区别在意思的主动与被动上。现在分词和过去分词除了自己的语法意义具有一般式和完成式之外，用作状语时，往往可以表示原因、方式、结果、条件、让步等意义；用作定语时，作后置定语的较多。这两种分词短语结构简练，经常用于科技文体中。从翻译角度讲，熟悉并掌握分词的译法，会使科技文章的译文更加短小精悍，内容更加生动和精确。

5.1.1　现在分词短语的译法

现在分词短语是英语特有的，它兼有动词、形容词、副词三项特征，在句中可置于句首、句中和句尾，可作定语、状语、表语、主语补足语等。由于英汉两种语言表达方式的不同，如果机械地把现在分词短语逐字翻译，将会使人感到生硬晦涩，不能确切地表达出英语原意。因此，在翻译过程中应该注意对现在分词短语灵活、恰当地翻译，保证原意的流畅。

5.1.1.1　现在分词短语作定语的译法

英语中的现在分词作定语可以前置也可以后置，但现在分词短语作定语时，一定置于被修饰成份的后面，构成后置定语。大量使用后置定语是科技文章的特点之一，其作用相当于一个定语从句，修饰前面的名词。因此，与定语从句一样，作定语的现在分词短语通常译成前置的名词修饰语，相当于汉语的“的”字结构。

例1　This is because a conductor carrying a current is surrounded by a magnetic field.

【译文】这是因为载流导体周围有磁场。

例2　Brass is an alloy containing a large proportion of copper.

【译文】黄铜是一种含有大量铜的合金。

例3　Sounds having the same frequency are in resonance.

【译文】具有相同频率的声音会共振。

例1~3中的现在分词短语 carrying a current、containing a large proportion of copper、having the same frequency 在句中后置分别修饰先行词 a conductor、an alloy、sounds。汉译时翻译成“的”字或省略“的”字的前置定语，更符合汉语的表达习惯。

5.1.1.2　现在分词短语作状语的译法

一般来说，尽管分词在句中不能单独作谓语，没有语法上的主语，但分词或分词短语在句中作状语时，与句子主语的逻辑关系是行为与行为主体的关系，或者说是逻辑谓语与逻辑主语的关系，即我们通常所说的分词或分词短语的逻辑主语应与句中主语保持一致。

现在分词作状语时，其功能相当于状语从句，表示动作或行为发生的时间、条件、原因、结果、让步、目的、方式或伴随等情况。翻译时需要根据上下文语义加入恰当的连词，使句子前后连贯，通顺易懂。

（1）时间状语

一般形式的分词短语指分词的动作一发生，谓语的动作便随即发生。翻译时可加入“……时”、“在……时候”、“当……时候”。

例 4　Flowing through a circuit, the current will lose part of its energy.

【译文】当电流流过电路时，要损耗掉一部分能量。

例 5　Using the electric energy, it is necessary to change its form.

【译文】使用电能时，得改变它的形式。

某些分词的一般式可将带-ing 的分词和主句的动词相比较，如果前者的动作没有结束，后者的动作就不能进行，表示“做了什么之后”的意思，则译为“……之后”。

例 6　Knowing current and resistance, we can find out voltage.

【译文】知道电流和电阻之后，我们就能求出电压。

例 7　Burning in oxygen or air, carbon forms into carbon dioxide.

【译文】在氧气或空气中燃烧之后，碳就转变成了二氧化碳。

完成形式的分词短语经常放在句首，指分词的动作完成后，谓语的动作才发生。翻译时可译为“在……之后”、“经过……之后”等。

例 8　Having obtained the initial conditions, we are going to solve the network differential equations.

【译文】在获得初始条件后，我们接下来就要解这些网络微分方程了。

例 9　The metal was hammer-hardened having been heated to a definite temperature in the furnace.

【译文】这个金属在炉子里加热到一定的温度后就被锤炼。

（2）条件状语

现在分词短语作条件状语时，多放在句首，通常可译成“如果……”、“假如……”、“一旦……就……”、“只要……”等。

例 10　Being no cause to change the motion, a body can move uniformly and in a straight line.

【译文】如果没有改变物体运动的原因，物体将匀速直线运动。

例 11　Using a transformer, power at low voltage can be transformed into power at high voltage.

【译文】如果使用变压器，低电压的电力就能转换成高电压的电力。

例 12　Solving (8), we have the following equation.

【译文】一旦解出（8）式，我们就可得出下列方程式。

（3）原因状语

现在分词短语作原因状语表示现在分词的动作是谓语动作产生的原因时，多置于句首，往往可增译“因为”、“由于”、“鉴于”等连词。为了使译文前后连贯、语气通顺，也可以在主句前面增译“故”、“所以”等词。

例 13 Possessing high conductivity of heat and electricity, aluminum finds wide application in industry.

【译文】由于铝具有高度的导热性及导电性，它在工业上获得了广泛应用。

现在分词短语作原因状语，根据上下文可不加任何连词。但是由于英汉语言的差异，在汉译时，要对 be 的现在分词 being 做出相应的汉语解释，故加入符合汉语表达的连接词“因为”、“由于……所以……”。

例 14 Being the most efficient, the largest power stations are running all the time at full load.

【译文】因为效率高，一些大型发电厂总是满负荷运行。

例 15 Being extremely light, water vapor often rises high in the sky.

【译文】由于水蒸气较轻，所以常常升入高空。

（4）结果状语

结果状语经常位于句末，其前用逗号分开。为了突出结果的意思，其前可有副词，翻译时可视情况添加“因而”、“因此”、“从而”、“所以”、“于是”、“便”、“就”等词，也可不加。

例 16 When iodine crystals are heated to 114 °C, they melt, forming liquid iodine.

【译文】碘晶体加热到 114 °C 时就融化了，从而形成液态碘。

例 17 The propeller of an airplane forces air backward, developing thrust.

【译文】飞机的螺旋桨迫使空气向后移动，于是就产生了推力。

例 18 The base and acid neutralize each other, forming a new substance.

【译文】碱、酸中和，形成了一种新物质。

（5）让步状语

现在分词短语作让步状语时，可根据上下文适当加入“虽然……但……”、“即使……也……”、“哪怕……也……”等。

例 19 Electronic computer having many advantages cannot carry out creative work and replace man.

【译文】即使电子计算机有很多优点，它们也不能进行创造性工作，不能代替人。

例 20 Not having been discovered, many laws of nature actually exist in nature.

【译文】虽然许多自然界的规律尚未发现，但它们确实存在于自然界中。

（6）方式或伴随状语

分词短语作方式状语或表示伴随情况时，所表示的都是比较次要的动作，是用来说明主要动作的，通常只用现在分词的一般式，另外所表示的动作和谓语动词所表示的动作差不多同时发生，因此翻译时一般不加任何词，直接译出即可。伴随情况多放在句尾，有时也可放在句首。

例 21 Diodes are like switches that open or close, depending upon the voltage applied.

【译文】二极管像是断开和闭合的开关，这取决于所加的电压。

原句中 depending on（upon）是科技文章中常见的表达方式，表示伴随的情况，可译为“与……”、“取决于……”、“定于……”。

例22　In sputtering，argon is commonly used，having the advantage that it is an atomic gas and chemically inert.

【译文】溅射中我们通常使用氩，优点是氩是一种原子气体，而且在化学性质上是惰性的。

例23　Even when we turn off the bedside lamp and are fast asleep，electricity is working for us，driving our refrigerators，heating our water，or keeping our rooms air-conditioned.

【译文】即使在我们关掉了床头灯深深地进入梦乡时，电仍在工作：为电冰箱提供动力，把水加热，或使室内空调机继续运转。

原句中的三个现在分词短语 driving our refrigerators、heating our water、keeping our rooms air-conditioned 在句中均作伴随状语，修饰谓语动词 is working。分词 driving 原意为“驾驶”，在此译为“开动”或“启动”。

5.1.1.3　悬垂分词的译法

分词短语在句中作状语时，其逻辑主语（亦称隐含主语）通常应是整个句子的主语；如果不是，而且其本身也不带自己的主语，就会被认为是一个语言失误。然而，这种不合逻辑的语言现象在科技英语中却很常见。这种分词叫做“悬垂分词”（dangling participle）或“无依附分词 ”（unattached participle）。这类现在分词短语的逻辑主语与主句的主语不一致的情况，用于进一步阐述对主句的观点态度或说明，汉译时可直接译出。

（1）悬垂分词的逻辑主语为不确定的概念

在泛指任何人时，现在分词（短语）的逻辑主语指的是不确切的或一般的概念，或含义模糊的概念（we，you）。

例24　Using similar techniques，the topic can be expressed in different lights.

【译文】（我们）即使使用同样的技术，也可用不同的见解描述这个主题。

例25　Operating with high phosphorus iron，the phosphorus content in the raw steel can be held to less than 0.2%.

【译文】（我们）用高磷生铁操作时，能将钢水含磷量保持在0.2%以下。

例24、例25中的现在分词短语 using... techniques、operating... iron 均在原句中作状语，但它们的逻辑主语都不是主句的主语，而是相当于 we 或 you 的逻辑主语。

（2）悬垂分词的逻辑主语为句中某个非主语成分

当句子的谓语动词不表示某一具体的行为，而是指客观存在时，在句中作状语的分词及其短语的逻辑主语常由句中其他成分（如名词普通格、物主代词、人称代词宾格等）表示出来。

例26　Being subject/Subjected to light，the conductivity of semiconductors is found to be high.

【译文】当半导体受光的作用时，我们发现其导电性很高。

现在分词短语 being... light 的逻辑主语是主句中的 semiconductors。

(3) 悬垂分词及其短语的逻辑主语是前面的整个句子

例 27 They increased the length of the wire, thus increasing its resistance.

【译文】他们把导线长度加长了，因而增大了导线电阻。

原句相当于 They increased the length of the wire, which/and this increased its resistance。现在分词短语 increasing... resistance 的主语为整个主句。

(4) 已转化为介词或连词的悬垂分词

有些表泛指的分词在长期使用中已逐步被当作介词或连词使用，其逻辑主语与句子的主语不一致。

例 28 Such discontinuities invalidate old paradigms, including those that may have been the basis of an electric plant's success.

【译文】这些不连续变化，使得旧的范式，包括那些以前一直是发电厂赖以成功的范式失去了效用。

原句中的分词 including 在译文中被译为“包括”，相当于英文介词 besides 的功能。

5.1.1.4 独立分词短语的译法

分词作状语时，其逻辑主语一般应与句子的主语一致。如果不一致的话，分词前面可以由一个名词或代词作为自己的逻辑主语，构成分词的独立结构，即独立主格结构，或称为带逻辑主语的分词结构。独立主格结构的位置相当灵活，可置于主句前、主句末或主句中，常由逗号将其与主句分开。分词的独立结构用于修饰整个主句，而不是主句中的一个词或词组。独立主格结构在科技文章中常用于表示时间、原因、条件和附加说明等。

例 29 An electron is about as large as a nucleus, its diameter being about 10^{-12} cm.

【译文】电子大约与原子核一样大，直径约为 10^{-12} 厘米。

例 30 The resistance being very high, the current in the circuit is low.

【译文】由于电阻很大，电路中通过的电流就小。

例 29 中的划线部分“直径……厘米”表示附加说明；例 30 中的划线部分“由于……很大”表示原因。

5.1.2 过去分词短语的译法

5.1.2.1 过去分词短语作定语的译法

(1) 译成“的”字结构的定语

过去分词作定语，通常放在所修饰的名词前，说明名词所处的状态；过去分词短语作定语，通常放在所修饰的名词后，强调被动的动作。在英译汉时，要按照汉语的表达习惯转换成前置的定语修饰语，译为“……的”。

例 31 The amount of heat given out by the heater increases in direct proportion to the length of time the heater is on.

【译文】加热器所放出的热量与加热器开的时间长短成正比。

原句中的过去分词短语 given... heater 译成汉语中具有主谓关系的“的”字结构的定语，此时要省略介词短语中的介词，将介词短语 by the heater 译为名词“加热器”。

例 32 An alloy is a substance composed of two or more metals fused together.

【译文】合金是由两种或两种以上熔合在一起的金属所组成的物质。

（2）译成“状语 + 动词 + 的”结构的定语

例 33　The typical problem of circuit and network theory is to determine the currents caused by the application of a given voltage to a given circuit or network.

【译文】电路和电网理论的典型问题是确定将一定电压加于一定电路或电网所引起的电流。

原文中的过去分词短语 caused... network 可译成汉语的“状语 + 动词 + 的”的语序结构。因此，汉译为“将一定电压加于一定电路或网络所引起的”，作前置定语修饰“电流”。

例 34　The principle and method taken in the experiment give us good message enlightenment.

【译文】在试验中采用的理论和方法给了我们一个很好的启示。

（3）“as + 过去分词短语”的译法

在科技英语中，常常遇见“as + 过去分词短语”的结构。英译汉时，要省译 as 后再译为“的”字结构的定语，并置于所修饰的名词前。

例 35　The major advantages of the transistor as used in electronic circuits are light weight, small space, low power consumption.

【译文】使用在电子电路中的晶体管的主要优点是重量轻，体积小，耗电量低。

例 36　In fact, materials as used in transistors do not exist in nature but are man-made.

【译文】实际上，晶体管中所用的材料是人造的，在自然界中并不存在。

5.1.2.2　译成并列分句或后续分句

例 37　A relay is a device that, based on information received from the power system, performs one or more switching actions.

【译文】继电器是一种装置，它依据系统信息，履行一个或多个开关作用。

原句中的过去分词短语 based... system 是句中定语从句 that performs... actions 的插入语，说明继电器履行开关作用时的依据是什么，汉译时可将 based 和 performs 两个动词看作并列的动词，译成相互平行的并列句来对主句作进一步的说明。

例 38　Solar power is significantly more expensive than coal or hydroelectric power, remained largely untapped.

【译文】太阳能发电比燃煤发电和水电都昂贵很多，因而仍基本上没有得到开发利用。

原句是因果关系，“由于昂贵，所以没有开发利用”，在翻译时既要体现出因果，又要表示出先后。所以，分词短语 remained largely untapped 译为主句的后续分句。

5.1.2.3　译成独立句

例 39　The energy source may be coal, gas, or oil, burned in a furnace to heat water and generate steam in a boiler.

【译文】能源可以是煤、油或天然气，它们在锅炉炉膛内燃烧从而将水加热成水蒸气。

原句中的过去分词短语 burned... boiler 是对主句介绍的三种能源功能方面的说明。为了使译文在形式上更加简洁，将过去分词短语表达的含义与主句一起译成了两个独立的句子，既符合汉语的表达习惯又完整地传达了原句的意思。

5.2 介词短语的翻译

英语中及物动词和介词后面都可以跟宾语。介词与其宾语一起构成介词短语。英语的介词短语在句中可以充当状语、定语、表语、宾语补足语和主语补足语等。

介词短语的翻译灵活、多样，对介词短语在句中作用的理解不同，所表达的意思也就不尽相同。因此，在翻译时必须根据上下文，对介词短语进行仔细推敲，并结合中文的表达习惯作结构上的调整，才能做到语言流畅，且不失原意。

5.2.1 介词短语做定语的译法

5.2.1.1 译为“的”字结构

介词短语在句中作定语，汉译时可以译为加“的”或不加“的”的定语结构，常置于中心词前面，而且英语的介词常被略去不译。

例 40 Depth of cut is actually the difference between the machined surface and the surface to be machined.

【译文】切削深度是加工的表面与待加工的表面之实际差值。

例 41 Increasing problems of all kinds with fossil fuel power plants are turning power and utility companies more and more to the use of nuclear powered electric generating plants.

【译文】矿物燃料发电厂所产生的问题日益增多，促使电力公司和公用事业公司越来越倾向于采用核动力发电厂。

原句中 problem... with 表示“在……方面产生的问题”，汉译时将 with 后面的词译为定语，with 按照汉语习惯被省译。

例 42 The electric resistance of a wire is the ratio of the potential difference between its two ends to the current in the wire.

【译文】导线的电阻等于该导线两端之间的电位差与导线中电流的比值。

5.2.1.2 译为动宾结构的句式

当由 of 引导的介词短语作定语（名词 + of + 名词），且 of 前的名词是动词派生词时，of 引导的介词短语译为动宾关系的句式结构更符合汉语的表达习惯。

例 43 The success rate of up to 90% claimed for lie detectors is misleadingly attractive.

【译文】据称，测谎器的成功率高达百分之九十，这颇有吸引力，但却容易把人引入歧途。

例 44 Two physicists discovered ceramics that could carry electric current without energy-wasting resistance at higher temperature than thought possible.

【译文】两位物理学家发现了一种陶瓷，这种陶瓷在高于预想的温度下，能够传输电流而不损失能量。

例 45 Physical chemistry applies the methods of physics to the study of chemical systems.

【译文】物理化学是把物理学的研究方法应用于研究化学规律。

上述例中，划线部分的介词短语在汉译时改变了结构，变为动宾关系，这种译文更容易被读者接受。

5.2.2　介词短语作状语的译法

5.2.2.1　译为状语从句

英语中介词短语作状语通常放在所修饰的词之后，而在译成汉语时，大多数置于它所修饰的动词、形容词之前，但也有放在后面的，视情况而定。

例 46　If an engineer applies a formula or a tabulation or a curve blindly, without knowing its history, its accuracy and the limits of its applicability, the results that he obtains will be neither intelligent nor practical.

【译文】如果一个工程师盲目地应用一个公式、一张图表或一条曲线，而对它的来历、精确度和应用范围却一无所知，那么他所得到的结果既不明智，也不实际。

例 47　The air is heated by the flame, expands, and thus becomes less dense. Denser air flows in from below and the result is that irregular layers of air of different densities are formed above the flame.

【译文】空气受火焰加热而膨胀，于是变得比较稀薄。较密的空气从底下流入，其结果是，在火焰上方形成了密度不同的不规则的空气层。

例 48　Alternating layers of semiconductors with slightly different compositions can act as mirrors, bouncing a portion of the light back and forth between the faces of the chip.

【译文】由于各层半导体的组分略有不同，所以能起到镜子的作用，使部分的光在芯片界面间来回反射。

上述例中，划线部分的介词短语分别译为条件状语、地点状语和原因状语。

例 49　Electrical power is always carried over long distances as a high-tension current at low-current strength, it is also sent as an alternating current, for such current is comparatively easily transformed either up or down in voltage.

【译文】电力总是以高电压低电流的方式进行远距离输送的，又是以交流电的形式送电的，因为这种交流电易于升压或降压。

原句中由 as 构成的两个介词短语均作方式方法状语，汉译时置于所修饰的动词之前。

5.2.2.2　译为句子的主语

介词短语在原句中作状语时，翻译时可译为汉语句子中的主语，这时英语的介词常被略去不译。

例 50　All of the atoms of an element contain the same number of protons in their nuclei, and the atoms of one element differ from those of all other elements in the number of protons so contained.

【译文】同一种元素的所有原子的原子核中含有的质子数都是相同的，而一种元素的原子与其他元素的原子的不同点就在于它们含有的质子数不同。

介词短语 in their nuclei 在原句中做地点状语译成句子的主语更符合汉语的表达习惯。

例 51　When ice changes into water, it does not expand. Instead, it becomes smaller in volume.

【译文】冰变成水时并不膨胀，相反，体积变小了。

原句中 in volume 这种表达度量的方式在英语中非常普遍，但汉语却没有这种用法，因此翻译时将其变为主语而意思不变。

5.2.2.3 译为并列分句

有些作状语的介词短语在汉译时，可以根据上下文转译为并列分句。

例 52 In the absence of mathematics' science would not exist.

【译文】没有数学，科学就不存在。

例 53 With the increase of pressure, the molecules get closer and closer.

【译文】随着压力增加，分子越来越接近。

原句中由介词 with 引导的介词短语所表达的概念与主句的意思是并列的，译成并列分句既符合汉语的表达习惯，又充分体现原句的含义。

例 54 This body of knowledge is customarily divided for convenience of study into the classifications: mechanics, heat, light, electricity and sound.

【译文】为了便于研究起见，通常将这门学科分为力学、热学、光学、电学和声学。

为了突出主句行为的目的性，汉译时将原句中的介词短语 for convenience of study 拆分出来，译为目的分句置于句首。

5.2.3 介词短语作宾语补足语的译法

作为宾语补足语的介词短语在汉译时，往往可译为谓语动词。

例 55 The theory was thought of considerable value for reference.

【译文】这项理论被认为有相当大的参考价值。

例 56 Heat sets these particles in random motion.

【译文】热量使这些粒子做随机运动。

例 55，例 56 中的介词短语结构是英语的常用方式，但在汉语中却比较少见。主要是因为英语使用名词性的结构较多，而汉语的动词则相对比较活跃，所以转译成谓语动词可以使句子更符合汉语的表达习惯。

例 57 A volt is the amount of electrical pressure required to cause a flow of 1 ampere of electricity through a conductor that has 1 ohm of resistance.

【译文】使 1 安培电流通过具有 1 欧姆电阻的导体所需的电压量称为 1 伏特。

原句中的介词短语 through... resistance 作宾语 flow 的补足语。

5.2.4 介词短语作主语补足语的译法

如果把带有宾语补足语的句子改为被动语态，则原来的宾语和宾语补足语就相应变为主语和主语补足语。在科技英语中，为了表示客观和谦虚的态度，或被动者比主动者更为重要，往往避免使用第一人称，而经常使用被动语态。

例 58 For a long time aluminum has been thought as an effective material for preventing metal corrosion.

【译文】长期以来，铝被看做是能够防止金属腐蚀的有效元素。

5.3 不定式短语的翻译

不定式是英语非限定动词的一种，是构成长句的一个重要因素。不定式有自己的宾语、状语和补足语，组成动词不定式短语。不定式短语可以起名词的作用，在句中做主语、表语、宾语、补足语等；也可以起形容词或副词的作用，在句中作定语或状语。因此，在翻译时要把不定式短语看做一个整体。

5.3.1　不定式短语作主语的译法

不定式短语作主语起名词、代词的作用，同时又具有很强的动词意义。不定式短语在句中的位置通常有两种：一种是位于句首；另外一种在正式的科技类文体中非常普遍，即当不定式结构较长时，要将其后置，并用引导词 it 作形式主语，在汉译时 it 可以省译。

5.3.1.1　译成动宾结构短语

例 59　To determine the optimum parameters will be time consuming.

【译文】确定最佳参数将是耗费时间的事。

例 60　Since we have already associated flow of charge or current with the existence of potential difference, it becomes of interest and importance to establish a quantitative relation between these two quantities.

【译文】因为我们已经把电荷或电流的流动与电位差的存在联系起来，在这两个量之间确立一种数量关系就成为重要而值得注意的问题了。

原句中的不定式短语 to establish... quantities 是真正的主语，it 是形式主语。汉译时要把不定式前置并保留不定式的动词意义，翻译成动宾结构。

5.3.1.2　译成介宾结构短语

当不定式短语常译为"将……"、"把……"、"用……"等介宾结构的短语，在汉语句子中通常做主语或宾语。

例 61　It takes a definite amount of heat to change a liquid into its gaseous state.

【译文】把液体变成气态要花费一定的热量。

例 62　It is possible for us to distinguish one material from another by the melting temperature of solids.

【译文】根据固体融化温度的不同就可以将它们彼此辨别开。

上述例中的不定式短语均汉译为介宾结构的短语，在汉语句子中分别作主语和宾语。

5.3.1.3　顺译与倒译

不定式短语在系表结构的句子中作主语或者是形式主语 it 的真正主语时，通常既可以采用顺译法，也可以采用倒译法。但有些情况下，只有其中一种译法复合汉语的表达习惯。

例 63　It is a handy method to test the transistor with an ohmmeter.

【译文1】简便的方法就是用欧姆表测试晶体管。

【译文2】用欧姆表测试晶体管是一种简便的方法。

原句中的不定式短语 to... ohmmeter 是形式主语 it 的真正主语，是对表语 method 的具体陈述。译文1为顺译，译文2为倒译，都可以准确地表达原意。

例 64　It is necessary to concentrate the minerals to be leached.

【译文】有必要对有待浸出的矿物进行选别。

原句中的 it 是形式主语，不定式短语 to... leached 是句子真正的主语。为了符合汉语的表达习惯，采用顺译法将原句译成了祈使句。

5.3.2 不定式短语作宾语的译法

不定式结构在科技英语中大部分都是用来充当宾语的，动词 want、decide、prefer、try、continue、begin、start 等后面常用不定式作宾语。当不定式短语在句中作宾语时，由于句子一般偏长，所以经常用引导词 it 作句子的形式宾语。汉译时，形式宾语不用翻译出来。

5.3.2.1 译成动宾结构短语

例 65 Radio continues to find wider application in science.

【译文】无线电在科学上继续获得日益广泛的应用。

例 66 An electric current begins to flow through a coil, which is connected across a charged condenser.

【译文】如果线圈同充电的电容器相连接，电流就开始流过线圈。

上述例中，不定式短语 to... application、to... coil 在原句中作宾语，分别译为动宾结构短语“获得日益广泛的应用”和“流过线圈”。

5.3.2.2 译成介宾结构短语

例 67 A machine is just a mechanical device which makes it possible to do work more conveniently by changing the applied force in directions or in magnitude or both.

【译文】机器只不过是这样的机械装置，它通过改变作用力的大小或方向，或既改变其大小又改变其方向的方法，使人们能够更加便利地做工。

原句中 it 是形式宾语，to do work more conveniently 是真正的宾语。汉译时，形式宾语不必翻译出来，不定式短语译为介宾结构短语。

5.3.3 不定式短语作表语的译法

不定式短语作表语时，可以译为“是……”或“就是……”的汉语句型结构。但在科技类文章中，动词不定式充当表语的情况比较复杂，翻译时要根据前后文的意思，依据具体情况进行处理。

5.3.3.1 译成动宾结构短语

例 68 The basic action of an SCR is to switch power on very rapidly.

【译文】可控硅的基本作用是迅速接通电力。

例 69 The function of a fuse is to protect a circuit.

【译文】保险丝的功用是保护电路。

5.3.3.2 译成介宾结构短语

例 70 Arc welding is to make metals together by means of an electric current.

【译文】电弧焊是利用电流将金属熔接在一起。

例 71 The two contacts at the base of the lamp are to carry current from the lamp holder.

【译文】灯泡底座的两个接触头通过灯座以传送电流。

上述例中的不定式短语在原句中作表语，汉译时采用了顺译法，不定式短语在译文中也作表语。

5.3.4 不定式短语作定语的译法

不定式短语作定语在英语中的使用非常广泛，它通常放在所修饰的名词之后，而在汉译时一般将整个不定式短语作为附加成分置于所修饰的名词之前，但有时也放在后面。当不定式短语作定语时与其所修饰的名词有三种逻辑关系：主谓关系、动宾关系和所属关系，了解这几种关系，能更好地理解和翻译句子。

5.3.4.1 译成主谓关系短语

当不定式的逻辑主语与不定式之间形成主谓关系，即逻辑主语是动作的执行者时，经常译成汉语的主谓关系短语。

例72 There was simply no visible evidence to support such a theory.

【译文】根本看不到有任何明显的证据来支持这一理论。

原句中 evidence 是不定式短语 to support such a theory 的动作发出者，它们之间是主谓关系，翻译的时候要先弄清这种逻辑关系，避免出现翻译错误。

例73 When the effect of inductance is such as to cause an induced voltage in the same circuit in which the changing current is flowing, the term self-induction is applied to the phenomenon.

【译文】当电感效应在变化的电流流过的同一电路里足以引起感应电压时，我们就用"自感"这个词来描述这一现象。

原句中 the effect of inductance 作时间状语从句中的主语，它是不定式短语 to cause an induced voltage 动作的发出者，两者之间是主谓关系。

5.3.4.2 译成动宾关系短语

动宾关系是指所修饰名词是不定式短语动作的承受者。如果不定式动词为不及物动词，后面应有必要的介词。

例74 In the design, there are many difficulties to overcome.

【译文】在这项设计中，要克服许多困难。

原句中 difficulties 是 to overcome 的宾语，两者构成汉语中的动宾关系"克服困难"。

例75 The cutting tool must be harder than the material to be cut.

【译文】刀具必须比所要切削的材料硬。

原句中的不定式短语 to be cut 为被动态，在译文里也应做出适当表达。

5.3.4.3 译成所属关系短语

当不定式与所修饰名词之间具有一种所属关系时，经常译成汉语中的所属关系短语，可以形成这种关系的名词有 way、time、plan、right、chance、opportunity、movement、tendency、reason、promise、wish、effort 等。

例76 In view of the drive to save energy, the weight reduction program of automobile is now actively pursued.

【译文】由于节约能源的压力，现正积极进行减轻汽车重量的研究。

例77 The best method to release stress is to anneal weldments.

【译文】解除焊接应力的最好方法是退火。

例78 Magnesium's tendency to vaporize can be reduced by alloying the metal with nickel or manganese.

【译文】镁的挥发倾向能通过把它同镍或锰作成合金而加以减小。

例76～例78中，三个不定式结构在句中均作后置定语，与所修饰的名词drive、method、tendency构成所属关系。

5.3.4.4 在there be句型中的译法

在there be句型中，用来作定语并修饰主语的不定式，常常译为汉语句子中的谓语。

例79 There are some metals to possess the power to conduct electricity and the ability to be magnetized.

【译文】金属具有导电能力和被磁化的能力。

原句中的不定式to possess修饰主语metals，译为汉语句子中的谓语。

5.3.5 不定式短语作状语的译法

在句子中，不定式短语有时用作状语，既可以对句中的动词、形容词或者副词进行修饰，也可以修饰全句。根据用法的不同，分别表示目的、结果、方式、原因、程度等。

5.3.5.1 作目的状语的译法

不定式作目的状语在句中的位置既可以放在句首，也可以放在句中或者句末。在汉译时通常采用顺译法并增译汉语中表示目的的词语，将“为了……”、“要……”等短语结构置于句首，将“以……”、“以使……”、“用来……”等短语结构置于句末。

例80 To extend this work and to investigate alternative processes, a joint project between DSIR and Carpentaria Exploration was initiated.

【译文】为了扩大这项工作和研究其他工艺方案，已开始执行科学与工业研究局同卡里宾塔里亚勘探公司的联合计划。

例81 To get a great amount of water power, we need large pressure and current.

【译文】要得到巨大的水力，就需要高水压和大流量。

在例80、例81中，不定式短语均作目的状语，翻译时按照汉语的习惯将其置于句首。

例82 They introduced new technological processes to raise their production.

【译文】他们采用新的技术工序，以便提高生产。

原句中的不定式短语to... production在汉译时被分解成了独立的目的状语分句，置于句末。

5.3.5.2 作结果状语的译法

不定式短语作结果状语通常位于句尾，一般采用顺译法进行翻译。在汉译时需要增译表示结果的词语“从而”、“使得”、“以”等。

例83 Methane unites with oxygen to yield carbon dioxide and water.

【译文】甲烷与氧结合，从而产生二氧化碳和水。

例84 A resistor is placed parallel with another only to make the current greater.

【译文】一只电阻器与另一只电阻器并联，从而使电流变大。

例85 When the effect of inductance is such as to cause an induced voltage in the same circuit in which the changing current is flowing, the term self-induction is applied to the phenomenon.

【译文】当电感效应在变化的电流流过的同一电路里足以引起感应电压时，我们就用“自感”这个词来描述这一现象。

上述例中的不定式短语均作结果状语，在翻译成汉语时增添了表示结果的词语“从而”、“以”。

5.3.5.3　在系表结构中作状语的译法

用作表语的形容词或过去分词后面往往可以跟不定式来说明产生这种结果的原因。汉译时可根据情况选择倒译或顺译。

例86　Loss of control is most likely to occur on inductive loads.

【译文】这种失控现象在电感负载时最可能发生。

不定式短语在原句中表示产生结果的原因，为符合汉语表达习惯以及原文的真实意思，翻译时无需使用表示原因的词语。

例87　These stresses are liable to occur through hammering or working the metal, or through rapid cooling.

【译文】由于锻造或加工，或由于迅速冷却，(金属中) 很容易产生这些内应力。

原句在汉译时增添了表示原因的词汇“由于”，使得译文更通顺。

例88　Up to now, copper alloys, according to CDA, have been more expensive to pressure die cast than aluminum and zinc, and a 25% reduction in the cost of pressure die casting of copper alloys is expected with the new technique.

【译文】按照英国铜业发展协会的意见，铜合金在压力模铸方面的费用迄今一直是高于铝和锌的，而在采用这项新技术后预计会降低25%。

如果用作表语的形容词表示性质或特征，则不定式短语是用来表示在哪方面体现了那种性质或特征，因此原句中的不定式短语译为“在压力模铸方面的费用”。

例89　Circuit breakers are necessary to deenergize equipment either for normal operation or on the occurrence of short circuits.

【译文】无论在系统正常运行或发生短路时都需要使用断路器来断电。

原句在汉译时使用了倒译的方法，将either...or引导的并列结构前置译为汉语里的时间状语从句。原句中的不定式短语放在句末与逻辑主语构成主谓结构。

5.3.6　不定式短语作补足语的译法

5.3.6.1　作宾语补足语

为了使意思相对完整，一些及物动词除要求接宾语外，有时还需要接宾语补足语来说明宾语的行为、状态、特征。不定式短语作宾语补足语时，句子的宾语就是不定式短语的逻辑主语，汉译时，可以顺译为主谓结构作整个句子的宾语。

例90　We consider heat to be a form of energy.

【译文】我们认为热是一种能量形式。

原句中的宾语是heat，不定式作宾语补足语，补充说明宾语的属性。

例91　Conductors allow electricity to pass through more or less freely.

【译文】导体可使电较为容易地流过。

原句中的宾语是electricity，不定式作宾语补足语，补充说明宾语的状态。

例 92 It is to be emphasized that a source of electricity current is simply a device for causing electricity to move around a circuit.

【译文】必须着重指出，电源只不过是使电流沿着电路流动的一种装置。

原句中 it 是形式主语，替代 that 引出的主语从句。不定式短语 to move... circuit 作宾语 electricity 的补足语。

然而，当有些特定的词连接不定式作宾语补足语时，往往省略 to，这类词包括 make（使得）、have（使）、let（让）和表示感觉的动词 see（看）、notice（注意）、watch（注视）、observe（观察）、hear（听）、feel（觉得）等。

例 93 Electricity makes a motor run.

【译文】电使电机运转。

例 94 Let F represent force.

【译文】设 F 表示力。

5.3.6.2 作主语补足语

不定式短语在句中作宾语补足语的句子，有时可以变为被动语态，原来担任宾语补足语的不定式就变成主语补足语，和主语构成一种逻辑上的主谓关系。汉译时，习惯将英语的被动句译为汉语的主动句或无主句，因此英语被动句中的主语又被译为宾语，而作主语补足语的不定式短语相应地转译为宾语的补语。

例 95 Heat is considered to be a form of energy.

【译文】我们认为热是一种能量形式。

例 96 The thermal decomposition of ammonium carbamate can be made to occur by the following methods.

【译文】氨基甲酸铵的热解可利用下列方法加以实现。

上述例中的被动句在汉译时转译成了主动句，而不定式短语译成了主语的补语。例 95 按照汉语习惯增译了主语“我们”。

例 97 Insulators are used to confine a current to the desired path.

【译文】绝缘体用来把电流限制在所要求的路线中。

虽然原句没有转换成汉语里的主动句，但不定式短语也相应地转为汉语的“把……”句型，与主语仍然保持一种逻辑关系。

5.4 同位语短语的翻译

同位语在科技英语中的运用尤为广泛。同位语常常用名词、代词、数词、形容词、不定式或短语来表示，主要出现在名词或代词之后，用以对其前边的名词或代词（亦可称为先行词）作进一步的解释、说明，以及阐述前者的情况、身份和特征等。在语法结构上，同位语与它所修饰的名词或代词处于同一地位，所以又称之为特殊形式的定语。

翻译时，若同位语成分较短，可以直接在插入位置翻译或者置于先行词前作定语。对于较长的同位语成分，一般应该处理成一个独立的简单句。

5.4.1 译成独立简单句

例 98 Everywhere there are irrigation projects: dams, reservoirs and canals-large and small, complicated or still under construction.

【译文】到处都是灌溉工程，有水坝、水库和水渠——有大的、有小的、有完工的、有正在修建中的。

原句中的同位语成分较长，表达的含义也较多，因而译为简单句更符合汉语的表达习惯。

例 99　The modern computer can even type letters you dictate to it, and that in any language you choose.

【译文】现代计算机甚至还能编辑出你口述的信件，并且用你所选择的任何语言编辑。

原句中的 and 引导同位语，that 替代了前面一整句话。一般来说，当同位语不是重复其中心词，而是用 that 或 one 来替代时，that 可以替代前面一整句话或部分内容，one 则替代表示具体事物的名词，而且这种同位语经常由 and 引入。

5.4.2　译成前置定语

例 100　Recent research shows that even moderate temperature sources can operate turbines using a substance rather than water——one with a lower boiling point.

【译文】新近的研究表明，如果不是用水，而是用沸点较低的物质，甚至中等温度的热源也能带动汽轮机。

同位语经常使用 one 来替代某一表示具体事物的名词。这时，同位语必须附加修饰语加以扩张词义，从而达到强调、精确、修正前面所述等目的。原句中用 one 替代 substance 一词，同时用 with 短语加以修饰，翻译的时候译为前置定语。

例 101　Archimedes first discovered the principle of displacement of water by solid bodies.

【译文】阿基米德最先发现固体排水的原理。

原句中 of displacement of water by solid bodies 是名词化结构，作 principle 的同位语，汉译时放在中心词的前面作定语。

5.4.3　直接在插入位置翻译

例 102　Most corrosion, especially (corrosion) of metals, is caused by oxidation.

【译文】大多数腐蚀，尤其是金属的腐蚀，是由氧化引起的。

由 especially、particularly、in particular、chiefly、mainly 等副词引出的同位语表示强调，这时同位语的中心词常省略，但在翻译时需要译出省略的中心词并且通常直接在其插入位置翻译。

例 103　The simplest atom, the hydrogen atom, contains one proton and one electron.

【译文】最简单的原子，即氢原子，含有一个质子和电子。

原句中 the hydrogen atom 作句子主语 atom 的同位语，形式上用逗号隔开，翻译时增加汉语中表示同位关系的词"即"。

5.5　名词短语的翻译

名词短语是以名词为主导词的短语，或者前面冠以其他词，如冠词、形容词或另外一个名词，或者后面接介词短语。名词短语在翻译时，为了使译文不显唐突、易于理解，应注意以下三种译法：

5.5.1 译成动宾关系短语

例 104 The application of just a fraction of a volt across the ends of a copper conductor can produce a large current flow.

【译文】在一个铜导体的两端只要加上零点几伏的电压就会产生一个相当大的电流。

原句中名词短语的主导词是由及物动词转化而来的名词，与 of 引导的介词词组构成逻辑上的动宾关系，译成“加上零点几伏的电压”，并与其他划线部分一起译成句子的主语。

5.5.2 译成主谓关系短语

主导词与介词短语中的名词也可能是主谓关系，如果主导词是从不及物动词转化过来的名词，后面仍接 of 短语，从逻辑上来理解，动作或行为正是由 of 后的名词发出的，它们之间的关系是主谓关系，后者是主，前者是谓。翻译时要把这个关系体现出来，即按主谓关系翻译。

例 105 The advance of science has been very great during the last fifty years.

【译文】近五十年来科学有了极大的进步。

例 106 Other requirements of the lathe tool are long life, low power consumption, and low cost.

【译文】车床的其他要求是使用寿命长，动力消耗少和造价低。

名词短语的最简单形式是“形容词 + 名词”，在翻译的时候比较灵活，可以把后面的名词看做主语，前面的形容词看做表语，构成主谓关系。原句中的 long life、low power、low cost 可以分别译为主谓关系短语“寿命长”、“动力消耗少”和“造价低”。

5.5.3 译成复合句中的分句

在理解名词短语的主导词与从属词之间关系的基础上，翻译的时候还要同上下文相呼应。为了清晰准确地表达原文信息，一些形式上较长的名词短语可译成复合句中的分句。

例 107 Remote control and supervision functions for line switches together with automatic sectionalization of faulty line section are provided.

【译文】在故障自动分段的基础上，还提供了遥控和监视功能。

原句中 together with 译作“在……基础上”更能清楚地表达原意，同时，由于原句中的名词短语较长，故汉译时分解了句子，使之成为复合句中的分句。

例 108 With loop arrangements all loads may be served even though one line section is removed from service.

【译文】用环网接线方式，即使一段导线从系统中被切除，仍不影响所有负载的供电。

原句中的“with + 名词短语”结构在汉译的时候分解成状语从句置于句首，突出强调原句所表达的意思。

5.6　形容词短语的翻译

形容词短语结构是英语中常见的一种语言现象。英语中某些形容词后面要求有一定的介词或不定式与之连用，这样就形成了形容词短语。形容词短语作定语时通常放在被修饰的名词之后，翻译时需要前置。

5.6.1　译成前置定语

例 109　In radiation, thermal energy is transformed into radiant energy, similar in nature to light.

【译文】热能在辐射时，转换成性质与光相似的辐射能。

例 110　Coal is a kind of fuel suitable for power generation.

【译文】煤是一种适合发电的燃料。

上述例中的形容词短语 similar in nature to light 和 suitable for power generation 分别用来修饰中心词 radiant energy 和 fuel，在原句中作后置定语，汉译时依照汉语表达习惯将其前置。

例 111　The gear teeth should have the strength sufficient to operate.

【译文】齿轮的齿必须具有足够的强度来运转。

sufficient to operate 是“形容词 + 不定式”构成的形容词短语，为准确清楚地表达原意，汉译时将形容词 sufficient 译为前置定语“足够的”，不定式 to operate 译为宾语的补足语。

5.6.2　译成并列式分句或后续分句

例 112　The machining process varies a lot, dependent upon the desired quality of the product and the nature of the material used.

【译文】机械加工程序变化很大，这取决于产品的期望质量以及所有材料的性质如何。

原句中的 dependent... used 是形容词短语，但由于结构较长，又是对先行词的解释说明，因而分译成了两个汉语句子。

5.6.3　译成句子的谓语

例 113　The fact that some bodies float on water and other liquids shows also that there exists a force acting against the lower surfaces sufficient to counteract their weight.

【译文】某些物体漂浮在水面上或其他液面上，这一现象也表明，存在着一个作用在物体底部表面的力，该力足以抵消物体的重量。

原句中，the fact that 引出同位语从句。there exists... 相当于 there be 的句型，这种句型都是倒装结构，主语是 force，谓语是 exists。分词短语 acting... weight 作定语，修饰 force。sufficient... weight 是形容词短语，也修饰 force。由于 force 分别由分词短语和形容词短语后置修饰，所以形容词短语译成独立分句作谓语。

5.7 多枝共干现象的翻译

5.7.1 多枝共干现象

“多枝共干”是科技英语中常出现的语言现象，即一个动词带有两个或多个宾语；两个或多个动词（短语）共用一个宾语；几个宾语共有一个动词；几个形容词、介词短语作定语共同修饰同一个名词；一个形容词或一个介词短语修饰几个名词；两个或多个动词共受一个状语的修饰，共有一个主语或宾语；两个或几个状语共同修饰一个动词等。由于这种语言现象犹如几根树枝长在同一根枝干上，因此被形象地称为“多枝共干结构”。

5.7.2 多枝共干结构的辨识方法与译法

在判断是否为“多枝共干”时，“多个名词（短语）+一个后置修饰语”的结构易引起误译。由于后置修饰语最多只能与一个名词（短语）临近，而与其他所修饰的中心词远离，译者在判断该结构中的后置修饰语是属于“多枝共干”中共用的成分，还是仅仅修饰最后一个名词（短语）时容易产生混淆。因此，必须掌握基本的辨识方法，在翻译时才能避免误译。

5.7.2.1 根据语法关系判断

根据语法知识来判断句中各成分的关系，是明确后置修饰语在句中的修饰关系的重要线索和依据。对于“多个名词+一个后置修饰语”是否为“多枝共干”结构，主要可根据以下三点加以判断：

（1）根据限定词判断

一般情况下，“后置修饰语”所指范围通常是在限定词之后，该后置修饰语之前。所以，若句中的多个名词（短语）共用一个限定词，如 the、a、all、any、every、other、some 等，一般可判断为“多枝共干”结构，否则就不是。

例 114 The machine is intended for grinding <u>the top and bottom surfaces of tungsten carbide tips</u>.

【译文】该机床是供磨削<u>硬质合金刀片上下表面</u>用的。

原句中的定冠词 the 同时修饰 top（surfaces）和 bottom surfaces，后置修饰语 of tungsten carbide tips 所指范围是 the top and bottom surfaces，故可以判断这是“多枝共干”结构。

例 115 Rods and wires shall have a smooth finish free from surface imperfections，corrosion products，grease or <u>other foreign matter which would affect the quality of the weld</u>.

【译文】焊条与焊丝应具平滑的表面粗糙度，不得有表面缺陷，腐蚀产物，油脂或<u>会影响焊接质量的其他杂质</u>。

other 在句中只修饰 foreign matter，故可以确定 which 引导的从句只修饰先行词 other foreign matter，原句并不是“多枝共干”结构。

（2）根据“一致原则”来判断

“一致原则”是指当后置修饰语为定语从句时，判断其是否为“多枝共干”结构所遵循的原则，即明确定语从句中的谓语动词与所修饰的中心词在数上是否一致性。

例 116　Some of the energy of the charge produces the sound and light which enable us to hear and see the spark.

【译文】电荷的部分能量产生了我们能够看到的火花，以及能够听到的火花的声音。

原句中的划线部分为"多枝共干"结构，因为定语从句中的谓语动词为复数，同时修饰 the sound and light。另外亦可根据冠词 the 同时修饰 sound and light 判断，定语从句的中心词为 the sound and light。

例 117　The purpose is to transmit only a chrominance signal for color and a luminance signal that contains the monochrome information.

【译文】其目的是仅仅传送一个色度信号和一个含有单色信息的亮度信号。

原句不是"多枝共干"结构，因为定语从句中的谓语动词为单数，不能修饰 and 连接的两个并列结构。

（3）根据"平行结构"来判断

"平行结构"是指由"等立连词"连接的两个或多个中心词在词性、结构、单复数等方面都是一致的，此时可判断句子属于"多枝共干"结构。

例 118　The physical dimension of the antenna determines the amount of inductance and capacity existing in the circuit and consequently the resonant frequency of the antenna system.

【译文】天线的尺寸决定电路中的电感量和电容量，从而也决定了天线系统的谐振频率。

原句是一个既含有名词与后置定语的"多枝共干"结构，又含有谓语与宾语"多枝共干"结构的句子。首先，原句中的谓语动词 determines 有两个宾语 the amount 和 the resonant frequency；其次，在宾语部分可以判断 inductance and capacity 是由等立连词 and 连接的平行结构，因此 existing in the circuit 既修饰 inductance 又修饰 capacity。

例 119　However，both the theory and the generation of FM are a good deal more complex to think about and visualize than those of AM.

【译文】然而，与 AM 相比，对 FM 的原理及其产生的理解和检测则更为复杂。

根据原句中的平行连接词 both... and 可以判断 FM 的中心词是 of 前的两个名词 theory 和 generation，因此属于"多枝共干"结构。

5.7.2.2　根据逻辑意义来判断

语义是语法的基础，仅仅根据语法知识来判断是不够的，有时甚至是不准确的。所以在利用语法知识的同时，一定要注意句子的逻辑意义以及科技意义，只有这样才能准确翻译。

在翻译带有"多枝共干"结构的英语句子时，首先要正确理解原文，准确分析和理解原句是翻译的前提和关键；其次，要根据英汉两种语言的差异，进行适当的转化；最后，根据汉语的表达习惯作适当的调整，使译文既准确表达出原文的含义，又通顺流畅。

例 120　The development greatly extends the range of applications and the reliability of the jet-type filter while retaining its other advantages.

【译文】这种改进大大地扩大了喷射式过滤器的应用范围，提高了它的可靠性，同时又保留了它的其他优点。

如果根据平行结构来判断，and 连接的两个名词词组 the range of applications 和 the reliability of the jet-type filter 同时作谓语动词 extends 的宾语。但是在分析句子的逻辑意义后可确定后置修饰语 of the jet-type filter 不但修饰 the reliability，还应该修饰 the range of applications。

以上论述了“多个名词（短语）+一个后置修饰语”的“多枝共干”结构的判断与翻译，另外还有五种结构同样属于“多枝共干”式结构，即多个定语（形容词或介词短语或名词）共同修饰同一名词；几个宾语共有一个动词或介词短语；一个宾语共有几个动词或介词短语；一个后置状语修饰几个动词；几个后置状语修饰同一个动词。

总之，多枝共干结构形式多样、用法特殊，如果出现在长句或段落里就更复杂。因此在判断一个句子是否有多枝共干结构，哪一个成分是共有成分时，要认真分析，仔细辨别，要根据上下文和全句所表达的意思做出正确的判断。有时除了上下文和句子意思外，有些语法特征，如定冠词、名词复数、词语的省略等，也有助于我们辨别某些词语是否属于多枝共干。此外，为了增强译文的表现形式，在翻译多枝共干中的共有成分时，要充分考虑到英汉两种语言的差异。

翻译练习

1. A material which has the property of elasticity will return to its original size and shape when the forces producing strain are removed.
2. Being cooled in the air, this kind of steel becomes harder and harder.
3. Having been well insulated, the wire may be used as a conductor.
4. Copper has less resistance than aluminum for the same size wire, but aluminum, being much lighter in weight, has less resistance per unit of weight.
5. The expanding gas rotates the blades of the turbine, thus giving up a large part of the power, which drives the compressor.
6. Some alloying elements make the grain of steel finer, thus increasing the hardness and strength of steel.
7. The photon, not having a material mass, maybe considered as having radiationmass.
8. Pure iron is a silver-white metal, melting at 1535 °C.
9. Molecules vary considerable in size and weight, ranging from the tiniest macromolecules to the largest macromolecules.
10. Judging by its conductivity, the metal may be aluminum.
11. A valence electron can move away from its atom, leaving it a hole.
12. An object at rest has no kinetic energy, its velocity being zero.
13. An electric current decomposes water into hydrogen and oxygen, hydrogen being liberated at the cathode.
14. We can store electrical energy in two metal plates separated by an insulating medium.
15. A ferrous metal is one chiefly made of iron.
16. With the last-mentioned approach, water as supplied under the existing head is passed through an ejector whose diffuser creates a vacuum.
17. Given the voltage and the current, the resistance can be determined according to Ohm's.
18. The volume of a given weight of gas varies directly as the absolute temperature, provided the pressure does not change.

19. The electronic microscope possesses very high resolving power compared with the optical microscope.

20. The energy in coal and oil came from the sun, stored there by the plants of millions of years ago.

21. This atom, with 92 electrons spinning about the closely packed and complex nucleus, is the element uranium, which seems to be the most complicated of the natural atoms.

22. Hysteresis energy is a useful basis for the establishment of failure criteria in fatigue.

23. For about the last 200 years, the structural theory has been developed by scientists, mathematicians and engineers, with the very practical object of providing a reliable basis for the design of structures by calculation.

24. With the base grounded, transistor Q4 is a very high impedance.

25. As an alternative to impedance functions, one can use a friction factor lookup table that relates the impedance between zones to the attractiveness between zones.

26. Some features of the nucleon structure emerge from these relations.

27. It is necessary to prevent induction motors against overload and under voltage.

28. The porous wall acts as a kind of seine for separating molecules.

29. To transmit electromagnetic waves takes energy.

30. To ensure that the reaction progresses reasonably fast involves the closest contact between the small account of liquid phase and the much larger bulk of solid phase.

31. It is rather difficult to define this quantity precisely.

32. It is essential to earth the rotor shaft by an additional earthed brush.

33. When oxygen is blown into molten pig iron, silicon begins to oxidize first.

34. The use of parallel circuits makes it possible to switch any light or appliance on or off without affecting other circuits.

35. Insulation is to keep electrical conductors separated from each other and from other nearby objects.

36. The natural tendency of heat to flow from a higher to a lower temperature makes it possible for a heat engine to transform heat into work.

37. To allow our silicon crystal to conduct electricity, we must find a way to allow some electrons to move from place to place within the crystal, in spite of the covalent bonds between atoms.

38. The capacity of individual generators is larger and larger to satisfy the increasing demand of electric power.

39. Though only 107 elements have been discovered, they combine in enough different ways to make the millions and millions of different things in the world.

40. In communications, the problem of electronics is how to convey information from one place to another.

41. Short circuits would cause very strong currents to flow.

42. Power makes machines run.

43. Friction compels a body in motion to stop.

44. It was not until the 19th century, when magnetism was shown to be the product of electricity, that serious research into the nature of permanent magnetism began.

45. Insulators are used to confine a current to the desired path.

46. A minimum electric power system is shown in Fig. 1, a combination of an energy source, a prime mover, a generator, and a load.

47. The fact that electric currents passing through a wire produce heat is known to all.

48. But electricity energy travels with the speed of light, about 186,000 miles in a second.

49. The creation of synthetic protein from petroleum to make food comes first on the list of current oil company research projects in come countries.

50. The primary concern of electrical engineering is the doing of work by the delivery of energy.

51. In these installations it is possible to increase Reynolds numbers by 3 ~ 9 times without further increase of the velocity head and the drive power.

52. Magnets have characteristics similar to electric charges in that like magnetic Doles repel each other and unlike magnetic Doles attract each other.

53. Without forces sufficient to overcome the resistance, bodies at rest will never move.

54. It was postulated in the early days of the subject that electricity was a fluid or rather two fluids, present in equal quantities inside matter and that charging a body consisted of adding an excess of positive of negative fluid to it.

55. The work piece will pass through a series of rollers, each a little closer to the next one.

56. The ultrasonic metal inspection is the application of ultrasonic vibrations to materials with elastic properties and the observation of the resulting action of the vibrations in the materials.

57. The rated current must be considered by dividing into pulse current (peak current) and continuous current which depend on the breakdown mode.

58. They will in general be divided in to a reflected train and a transmitted train whose relative intensities will depend on the magnitude of the velocity changed at the boundary, on the abruptness of this change, and on the angle of incidence.

59. Surface wear can be caused by either thermal stress or fatigue in the sliding faces.

60. It has been shown that water in the fluid can affect both the fluid integrity and the operational behavior of the system components.

第6章　句子的翻译

6.1　句子翻译的基本步骤

科技英语句子翻译的全过程可分为两个阶段：原文理解、译文表达。理解是表达的前提，若不能正确地理解原文就谈不上忠实准确地表达，但理解与表达通常是互相联系的过程，在进行汉语表达的时候，又可以进一步加深对原文的理解。

6.1.1　原文理解

原文理解阶段的目的在于读懂英语原文，弄清原文的意思，理解各部分之间的语法与逻辑关系。这一阶段主要分为提炼主干、句子成分分析和辨别词义几个步骤。由于前面章节中对词义的辨别已经详细阐述过了，在此不再重复。在这一部分，我们主要探讨一下提炼主干和分析句子成分这两个环节。

6.1.1.1　提炼主干

在原文理解阶段，应注意首先提炼主干，即找出句子中的主、谓、宾、表等主干成分，从而凸现句子的主体框架，为理解句子的完整涵义打下基础。

例1　Ice is frozen water or water that has become solid.

【译文】冰是冻结的水或成为固态的水。

英语中，一个句子只能有一个谓语或几个并列的谓语，由此，句子中谓语的数量是判断句子个数的标准。例1中有多个谓语，但经过提炼主干，其主体框架就很明显了：Ice is ... water or water that...，所以其骨架涵义为“冰是一种……的水”。

6.1.1.2　分析句子成分

分析句子成分是指在弄清句子的主干后，还要识别句中各个成分的作用及语法关系；在复合句中，要根据从属连词、并列连词、关系代词或关系副词等理清主句和从句及其语法和逻辑关系。

例2　The diode consists of a tungsten filament, which gives off electrons when it is heated, and a plate toward which the electrons migrate when the field is in the right direction.

【译文】这种二极管由一根钨丝和极板组成。钨丝发热时放出电子，并在电场的作用下向极板运动。

原句中有5个谓语动词：consist of、give off、is heated、migrate、is...，可以知道这个句子是复合句。主干是 diode consists of... filament... and... plate；两个 which 分别引导两个非限制性定语从句分别修饰 filament 和 electrons；在两个定语从句中又分别嵌有两个由 when 引导的状语从句，这两个从句都表示条件。

6.1.2　译文表达

表达就是译者把自己对英语原文理解的内容用汉语表达出来，表达的好坏取决于译者对于英语原文的理解程度以及汉语水平。理解是表达的基础，表达是理解的结果，但是正确的理解并不意味着一定会有正确的表达，因为在表达上还存在许多具体的方法和技巧。在此我们只介绍两种基本的翻译方法，直译和意译。

（1）直译

所谓直译，就是在译文语言条件许可时，在译文中既保持原文的内容，又保持原文的形式。

例 3 The function of a fuse is to protect a circuit.

【译文】保险丝的功用是保护电路。

这个句子是典型的"主—系—表"结构，而且几乎没什么修饰语，与目的语汉语的句子成分语序基本一致，因此可以采取直译法，将各部分词义一一对应即可。

（2）意译

汉语和英语分别属于不同的语系，两者在词汇、句法结构和表达方法上具有很多的差异。当原文的思想内容与译文的表达形式有矛盾不易采用直译的方法处理时，就应采用意译法。

例 4 It goes without saying that these waves travel at the speed of light.

【译文】显然，这些波是以光速传播的。

在科技英语中经常用 it 作形式主语，而把真正的主语从句或不定式等其他非谓语所充当的主语放在后面。这种结构往往谓语较简短，可以避免头重脚轻。但汉语没有这样的结构，为了能够准确表达原文意思及保证译文的通顺流畅，要把形式主语 it 和它的谓语译成一个短语，把从句译成句子的主干。

例 5 Let us develop an equivalent circuit for a two-wire power transmission line of length L, diagrammatically shown in Fig. 2.

【译文】下面推导如图 2 所示的一长度为 L 的双输电线的等值电路。

原句是一个祈使句，宾语 circuit 之后的部分起修饰或说明作用，汉语中这样的成分应该前置。但是，如果按照字面意思将 two-wire power transmission line 译为"双电线输电线"，将 diagrammatically shown in Fig. 2 译为"以图表形式出现在图 2 中的"，就会得出下列译文：让我们来推导一下以图表形式出现在图 2 中，长度为 L 的双线输电线的等值电路。很显然，第二个译文中语言表达拖沓冗长，术语表达不专业。按照汉语的习惯，一般在用图表表现数据时我们都说"如图……所示"；从原文的上下文来看，two-wire power transmission line 的对等汉译形式应该是"双输电线"。

6.2 简单句和并列句的翻译

6.2.1 句子成分的译法

英语中的主要句子成分有主语、谓语、宾语、表语、定语、状语、主语补足语和宾语补足语。因为英汉两种文化和语言表达习惯的差异，我们不能按照原文的遣词造句习惯来构架译文。尤其是科技英语的翻译，既要求译文能准确表达原句的内容、符合技术规范，又要通顺易懂、符合汉语的遣词造句习惯。这就要求译者在某些情况下，在理解原文的基础上，根据汉语的表达习惯对原句成分进行适当的转换。

6.2.1.1　主语的译法

（1）直译

当英语句子的主语是名词、代词、形式简单的动名词或不定式时，一般直译为汉语的主语。

例6　Installing the electronic units in the ground and leasing telephone wires would cost $15,000,000.

【译文】安装地下电子装置和租借电话线要耗资1500万美元。

例7　You may feel its electric charge flow through your body and away into the earth.

【译文】你会感到电流从你的身体中流过，然后离开而进入地下。

例8　Electrons revolve about the nucleus at tremendous speeds.

【译文】电子以极高的速度绕核旋转。

上述例子都是将原文中的主语直译成主语。

（2）转译为谓语

① 当英语句子的主语是具有动词意义的名词，如care、need、attention、thought、overview、attempt、emphasis等，或在there be句型中，或为系表结构，或在被动句中时，可以译成谓语。

例9　Stress must be laid on the development of the electronics industry.

【译文】（我们）必须重视电子工业的发展。

例10　Care should be taken at all times to protect the instrument from dust and damp.

【译文】应当始终注意保护仪器不能沾染灰尘和受潮。

例11　An overview of some latest breakthroughs in the controller design of multivariable linear time-invariant systems is made.

【译文】本文概括了近来在多变量线性时不变系统控制器设计中所取得的一些进展。

例9和例10译文较短，可以译为警示性无主句。例11译文较长，可视情况适当添加主语“本文”。上述例子中，stress、care、overview的汉语释义分别为“重视”、“注意”和“概括”，因为这些名词本身就具有动词性意义，因此转译成动词。

② 当原句主语是由动词或形容词派生而生成的名词时，可转译为动词。

例12　The elimination of certain problems of the generator is distinctly with the realm of possibility.

【译文】这种发电机的某些问题显然可以摒弃的。

例13　Growth in business involving advertising, accounting, and computer services has been especially rapid.

【译文】与业务有关的广告业、会计业以及计算机服务增长得尤其迅速。

例14　The lightness and cheapness of the substance recommended if for architectural decoration.

【译文】这种材料既质轻又便宜，建筑上常用它作装饰材料。

例11中的主语elimination衍生于动词eliminate，例12中的growth衍生于动词grow，这样的名词也具有动词意义，因此可以转译成谓语。科技英语中这类词还有improvement、application、development、transmission等。例13的主语lightness和cheapness分别衍生于形容词 light 和 cheap。

（3）转译为宾语

① 英语的被动句有时会被译成汉语的主动句，此时原句中的主语往往被译成宾语，一般需要把汉语译文处理成一个无主句，或在译文中采用泛指的人称作主语。

例 15 Unless a generator is provided, a battery system with automatic charging features should be provided.

【译文 1】如果没有发电机，则应配备有自动充电能力的蓄电池系统。

【译文 2】如果没有发电机，我们就需要配备有自动充电能力的蓄电池系统。

provide 和 generator、a battery system with automatic charging features 在逻辑上具有动宾关系。因此，汉译时把它们译成动宾关系的短语。由于没有明确指出施动者，译文中的主语可以省译或适当的采用泛指人称代词“我们”。有时也可增译其他表示泛指的主语“有人”、“人们”等。

例 16 Nuclear energy thermoelectric generators are proposed as power sources for the space explorers.

【译文】有人建议将核能温差电池用作太空探测器的能源。

② It be + p. p. + that clause 结构和 there be 句型在科技英语翻译时也常常被转换成无主句或增译泛指人称代词，这时一般将 that 引导的主语从句和 there be 句型中的主语转译成宾语。

例 17 There is the proposal that nuclear energy thermoelectric generators be used as power sources for the solar system deep space explorers.

【译文】有人建议将核能温差电池用作太阳系远程空间探测器的能源。

该句的主语是 proposal，其后 that 引导的是 proposal 的同位语从句，其作用是解释说明 proposal。因此，在翻译时可以增译主语“人们”或“有人”。

例 18 There are two imported 300MW generating sets in our power plant.

【译文】我们电厂进口了两台 300MW 的发电机组。

由于原句中提到了施动者 our power plant，可以将这个短语译为主语，而过去分词 imported和 300 MW generating sets 虽然在原句中是偏正关系的名词短语，但二者实质上是动宾关系。由于汉语中多用主动语态，因此，就将这个偏正关系的名词短语转译成汉语的动宾短语，其主语转译成宾语。

（4）转译为定语

为了符合汉语习惯，或使译文表述通畅，英语句子中的主语在译成汉语时，往往需要改译成定语。

① 在英语句子中，当句子的谓语被转译成汉语的主语时，常常将原句中的主语译成汉语句子的定语。

例 19 Electronic circuits work a thousand times as rapidly as nerve cells in the human brain.

【译文】电子电路的运行速度是人脑中的神经细胞的一千倍。

当把原句中的谓语 work 译成主语“运行速度”时，原句中的主语 electronic circuits 转译作定语“电子电路的”。

② 在 have 作谓语的句子中，have 的宾语往往被转译成汉语的主语。为了使译文更符合汉语的表达习惯，常常将原英语句子的主语转译成汉语句子的定语。

例 20　It is evident that semiconductors have a lesser conducting capacity than metals.

【译文】显然，半导体的导电性比金属差。

例 21　These electric machines have well-designed structures.

【译文】这些电机的结构设计得很好。

例 22　Ultraviolet rays have a larger amount of energy than visible light does.

【译文】紫外线的能量比可见光大。

以上三例中，have 后面的宾语要么是表示主语的性质，要么是陈述对象主语的组成部分，总之与主语有着密切的关系，这种情况下，往往把主语译成宾语的定语，而原句中宾语之前的定语一般根据情况转译为谓语“比……差”、“设计得好”和“比……大。

③ 在系表结构的句子中，当介词的宾语被转译成汉语的主语时，可以将原句中的主语转译成汉语句子中的定语。

例 23　This electric machine is simple in design, yet efficient in operation.

【译文】这台电机的结构简单，但效率很高。

例中，介词 in 的宾语 design 和 operation 分别译成主语，而主语 electric machine 转译为定语。

（5）转译成状语

在英语中表示时间、原因、条件含义的主语译成汉语时，一般需要转译成状语。

例 24　The acquaintance of science means mastering the law of nature.

【译文】如果懂得科学，就意味着掌握自然规律。

例 25　The application of the innovative design makes the electric machine become smaller and more efficient.

【译文】由于在设计方面的革新，电机的体积变小而且效率提高。

例 24 和例 25 中，由于主语 the acquaintance of science 和 the application of the innovative design 分别是 the electric machine become smaller and more efficient 和 mastering the law of nature 实现的原因和条件，因此可以将两个例句主语分别转译为条件状语“如果懂得科学”和原因状语“由于在设计方面的革新”。

例 26　The 20th century saw the rapid breakthroughs in the research of nuclear energy in China.

【译文】在 20 世纪，我国在核能研究方面取得了重大突破。

例 27　The recent decades has used digital computer techniques to improve the performance of more complex processes.

【译文】近几十年来，已运用数字计算机技术来提高更为复杂过程的运行性能。

如果按照字面意思直译，显然不符合汉语的表达习惯。分析一下句子各成分之间的关系可知，例 26 中，“重大突破”（rapid breakthroughs）的取得是发生在 20 世纪；例 27 中“用数字计算机技术来提高更为复杂过程的运行性能”发生在近几十年，因此，汉译时可将做主语的时间转译成状语。

6.2.1.2 谓语的译法

在翻译英文句子的谓语时，最常见、最普通的方法就是直译法。但有时也需要转译或省译。

（1）转译为主语

当英语原句中的谓语是表示状态的动词时，往往可以把作谓语的动词译成名词，在译文中作主语，而原文中的主语则一般转译成定语，一般译成“……的……是……”句型。

例 28 Gases differ from solids in that the former have greater compressibility than the latter.

【译文】气体和固体的区别在于前者比后者有更大的可压缩性。

例 29 Current is defined as the amount of electric charge flowing past a specified circuit point per unit time.

【译文】电流的定义是，单位时间内通过电路上某一确定点的电荷数。

例 30 This paper aims at discussing the application of solar energy.

【译文】本文的目的是讨论太阳能的应用。

例 31 Analogue computers are not as accurate as digital computers.

【译文】模拟计算机的准确性不如数字计算机。

例 28 ~ 31 中的谓语 differ、define、aim 和 are accurate 的意思分别是“与……不同”、“给……定义”、“以……为目的”和“准确”，都是表示状态的词。因此，可以分别译为“……的区别”、“……的定义是……”、“……的目的……”和“准确性”。

（2）转译为状语

例 32 The power station managed to maintain its supply of electricity day and night.

【译文】电站尽力保持日夜供电。

例 33 Under this circumstance, we usually tend to choose rubber, which is a better dielectric but a poorer insulator than air.

【译文】在这种情况下，我们往往倾向于橡胶，它的介电性比空气好，但绝缘性比空气差。

例 34 It concentrated on installing radio in ships.

【译文】它主要为船只安装无线电收发机。

例 35 Most U. S. spy satellites are designed to burn up in the earth atmosphere after completing their missions.

【译文】美国绝大多数间谍卫星，按其设计，就是在完成使命后，在大气层中焚毁。

在例 32 ~ 35 中，谓语 manage、tend、concentrated 和 designed 强调做事的状态、习惯和目的等，而且后面的动词不定式短语是谓语实质的表意部分，因此将谓语动词转译成状语“尽力”、“往往”、“主要”和“按其设计”。

（3）省译

当英语原句的谓语中含有系动词或者 prove to be、turn out to be 等动词时，在译成汉语时，往往可以省略不译。

例 36 The alpha rays proved to be charged electrically with a positive charge.

【译文】阿尔法射线原来是带有正电荷的。

上述句子中的 proved to be 没有译出。

6.2.1.3　表语的译法

英语句子中的表语可以由名词、代词、动名词、不定式等充当。表语有以下几种不同的译法：

（1）直译

直译就是按照原句子的系表结构的表达顺序翻译。

例 37　The function of amperemeter is to measure the amount of electric charge flowing past a specified circuit point per unit time.

【译文】电流表的功能就是测量单位时间内通过电路上某一确定点的电荷数。

原句的表语是不定式 to measure the amount of electric charge flowing past a specified circuit point per unit time（测量单位时间内通过电路上某一确定点的电荷数），由于句子是“主—系—表”结构，且主语很短，完全可以按照原句的顺序直译。

（2）译成汉语的主语

例 38　The rotor is a well-designed structure consisting of a laminated core containing a winding.

【译文】转子的结构设计得很好，他由一个叠片铁芯组成，铁芯上绕有线圈。

例 39　This metal is less hard than that one.

【译文】这种金属的硬度比那种差。

例 40　Zirconium is almost as strong as steel.

【译文】锆的强度几乎与钢一样。

英语句子的主语有时可以转译为汉语的定语，在这种情况下，原句中的表语往往转译为汉语的主语。例 38 在结构上是简单句，但表语是偏正关系的名词短语 well-designed structure，根据意义表达上的逻辑关系，在汉译时一般将这一部分译成汉语的主谓短语，在译文中充当一个小分句，表语的中心语 structure 转译为分句的主语，定语转译为小分句的谓语。若原句的表语为形容词，该形容词通常转译为名词。例 39 和例 40 中做表语的形容词通常表示状态、性质或特点，翻译时都转译成表示该性质和属性的名词，在译文中作主语。

（3）转译成汉语的谓语

当英语中作表语的是具有动词意义的名词时，可以将该表语转译为谓语。

例 41　These new machines are now in wide use.

【译文】这种新的机器现在广泛应用。

例 42　At present some old types of engines are still in use.

【译文】目前，某些老式发动机仍在使用。

例 43　Electronics is the study of the flow of electrons and the application of such knowledge to practical problems in communications and electric controls.

【译文】电子学研究电子运动规律，并把这种知识运用于通信和电气控制的实际问题。

当表语为“介词 + 具有动词意义的名词”结构，或当具有动词意义的名词后面有“of + 名词”作定语时，该结构中的名词与被这个 of 结构修饰的名词之间存在逻辑动宾关系，翻译时，一般都把这个结构同它所修饰的名词转译成动词短语。

（4）转译为宾语

英语常用系表结构来表达某物质的特性，表语常采用形容词或“of + 名词”。这时，一般把系动词译为“具有”、“属于”等动词，而表语则译为相应的名词，在译文中充当宾语。

例 44 The early steam engines were of the piston-type.

【译文】早期的蒸汽机属于活塞型。

例 45 The function of a motor is to convert electric energy into kinetic energy.

【译文】电动机的作用就是把电能转换成动能。

在例 44 中，表语中的介词 of 表示一种所属关系，因此在翻译时将 of the piston-type 处理成动宾关系的短语“属于活塞型”。例 45 中不定式短语作表语，英语中许多的“主—系—表”结构与汉语的判断句是对等的表达方式，这种情况下，英文中的表语同汉语判断句中的宾语是对等的。

（5）转译成状语

当英语的形容词作表语时，有时可转译成汉语的状语。

例 46 But in the case of electrical energy, the development from two completely chance observations were direct and very swift.

【译文】电能是从两项非常偶然的观察直接而迅速地发展起来的。

原句中的主语是具有动词意义的名词 development，这种情况下，主语需要转译成谓语，而对主语进行表述的表语就相应转译为作副词作状语来说明译文中的谓语。因此，将 direct 和 swift 转译成“直接而迅速地……”。

6.2.1.4 宾语的译法

（1）直译法

例 47 They are building a thermal power plant.

【译文】他们正在建一个热电厂。

主谓宾结构是汉语和英语中均存在的句型结构，汉译时往往可以直接对应：原文中的宾语在译文中仍然作宾语。

（2）转译为主语

由于思维方式和文化习俗上的原因，英汉两种语言在叙述事物的次序上存在差异。因此，英语句中的宾语往往要转译为主语。

① have 作谓语的句子往往把 have 的宾语译成主语。

例 48 The output of transformer has a voltage of 10 kilovolts.

【译文】变压器的输出电压为 10 kV.

例 49 As a working constant the PV factor has doubtful validity.

【译文】作为工作常数，PV 值的正确性令人怀疑。

例 50 This electrolysis system has an advantage in that no fossil fuel is used.

【译文】电解系统的优点在于其不使用矿物燃料。

例 48 ~50 中谓语 have 的宾语中心词分别为 voltage、validity 和 advantage，此时可以将该词转译作主语，原句的主语分别转译作定语“输出电压”、“PV 值的正确性”和“电解系统的优点”，而原句中宾语的定语分别转译为谓语“是 10 kV”或“为 10 kV”、“令人怀疑”、“在于其不使用矿物燃料”。

② 句中的介词短语译成主语，并省译介词。

例 51 The activity coefficient changes in value with increasing ionic strength.

【译文】活度系数的值随离子浓度的增大而增大。

例 52 Microwave electronic components and sub-systems are encountered during the development of systems such as those mentioned.

【译文】开发上述系统会遇到微波电子部件和子系统。

例 51、例 52 中的 in value 和 during the development of systems such as those mentioned 都是动词后面的状语，在汉译时分别转译成了主语“值”和“开发上述系统”。

③ 若英语句子的宾语是由动词派生而来，或是具有动词意义的名词时，常常译为汉语的谓语。若该宾语有形容词修饰，可以将该形容词转译为状语从而修饰谓语。

例 53 In the field of power engineering, we should make full use of the latest development in science and technology.

【译文】在电力工程领域，我们应该充分利用科技的最新发展。

例 53 中的宾语 use 具有动词意义，在汉译时可将其译为动词“利用”，宾语前的形容词 full 相应的转译成状语“充分”或“完全”来修饰谓语“利用”。

6.2.1.5 定语的译法

（1）顺译和倒译

定语的顺译就是按照英语词语的顺序将英文的定语译作汉语的定语。一般单个的词作定语时，采用这种译法。如：a steam boiler 蒸汽锅炉；a power station 发电站

当英语中的定语置于中心词后面时，汉译时一般采取倒译法。

例 54 The energy not consumed in the system will be stored up in the form of potential energy.

【译文】系统未消耗的能量将以势能的形式储存起来。

原句中的两个定语 not consumed in the system 和 of potential energy 分别放在 energy 和 form 的后面起限定修饰作用。汉语的限定修饰语习惯上放在所修饰词的前面，因此，汉译这两个名词短语时需要逆转定语和中心语的词序，分别译为“未消耗的能量”和“势能的形式”。

（2）转译

科技英语中常常将偏正关系的名词短语译成主谓短语，即英语原文的中心词译作汉语主谓短语中的陈述对象，定语转译成主谓短语中的逻辑谓语，整个短语翻译成了汉语的一个小分句。

例 55 Being a good insulator, rubber strongly resists current flow.

【译文】橡胶绝缘性很好，基本上能阻止电流通过。

例 55 中宾语的中心词 insulator 转译成了分句的主语，它的定语 good 转译成小分句的谓语，而且把具体名词 insulator，译成一个表示该事物特点和性能的抽象名词“绝缘性”。

① 转译为主语或中心词

如果主语或中心词后面有介词 of 引出的定语，汉译时常常把 of 引出的定语译成汉语的主语或短语结构中的中心词。

例 56 A minimum of 1.5 to 2.0 milligrams/litre of dissolved oxygen is maintained in the aeration tank.

【译文】充气槽内的溶解氧最少应保持在 1.5 ~2.0 毫克/升。

of dissolved oxygen 和 of electricity 分别作主语 A minimum of 1.5 to 2.0 milligrams/litre 和 kind 的定语，汉译时将 dissolved oxygen 和 electricity 转译成主语。

例 57 There are two kinds of electricity, which we call positive electricity and negative electricity.

【译文】电有两种：正电和负电。

② 转译为谓语

当句子谓语为 have，修饰宾语的定语为形容词时，为了使译文的表达更符合汉语习惯，通常将定语译成谓语。

例 58 The material exhibits a poor elasticity.

【译文】这种材料的弹性差。

例 59 Synchronous motor usually has a higher efficiency than that of a comparable induction motor.

【译文】通常，同步电动机的效率比相应的感应电动机的效率高。

例 60 Aluminum alloy has low specific electrical resistance and high thermal conductivity.

【译文】铝合金的电阻很低，而导热性很高。

例 58、例 59 中，谓语 exhibits 和 have 的宾语 elasticity 和 efficiency 同前面的定语higher 和 poor 分别转译成主谓关系“效率比……高”和“弹性差”。例 58 的主语转译成汉语句子中的主语的定语“这种材料的”。例 60 则将偏正名词短语译为分句，原句中 have 的宾语 electrical resistance 和 thermal conductivity 转译成主语，而它们前面的定语 low 和 high 相应转译成谓语，句子的主语转译成汉语句子中的主语的定语“铝合金的”。

在 there be 句型中，若主语后面有不定式短语或分词短语作定语时，可将其译成谓语。

例 61 There is not yet enough evidence to confirm the feasibility of the research.

【译文】还没有足够的证据证实这项研究是否可行。

例 62 There is a broad range of research interests in the department, including both traditional and emerging areas of electrical engineering.

【译文】该系的研究课题广泛，涉及电气工程的传统领域和新型领域。

例 63 There is a large amount of energy wasted due to friction.

【译文】摩擦损耗大量能量。

例 64 The water rushing down the pipe and hitting the blades sets the wheel turning.

【译文】水从管子里直冲下来，打在轮叶上，使轮子转动起来。

例 65 When you turn electric appliance off, there will be electric voltage built up in the switch, but no current will flow.

【译文】当关上电器时，在开关上有电压产生，但没有电流通过。

例 61 中，evidence 和 to confirm 是逻辑主谓关系，故将 to confirm 转译成句子的谓语。例 62 中 research interests 和 including 是逻辑主谓关系，在原句中 including 这个分词短语是 research interests 的定语，故将 including 转译成句子的谓语。例 63 中的 wasted 和例 64 中的

rushing down the pipe and hitting the blades 分别是主语的后置定语，翻译时都转译成了谓语"损耗"和"直冲下来，打在轮叶上"。例 65 中过去分词短语 built up 在句子中作 voltage 的定语，二者具有逻辑主谓关系，又因为 electric voltage 在 there be 句型中作主语，因此，将 built up 转译成句子的谓语。

③ 译为汉语的状语

当英语句子中具有动词意义的名词在汉译时转译成动词时，原句中修饰该名词的短语往往相应转译成副词，充当状语。

例 66　A superficial glance at the table shows a strong correlation between the two sets of measurements.

【译文】大致看一下这张表，就可以发现这两组测量数据密切相关。

例 67　To increase precision the machine tool must also provide rigid control of relative movement between the tool and work.

【译文】为了提高精密度，机床还应严格控制刀具与工件之间的相对运动。

例 68　A brief description of how transistors are manufactured is given here.

【译文】这里简单扼要地讲讲晶体管是如何制造的。

例 66 ~ 68 中的主语 glance、control 和 description 是具有动词意义的名词，因此，在汉译时要译成动词，原句中这三个词前面的定语需要转译成状语来修饰谓语。这样，例 66、例 67 由原句的一个句子转译成了两个具有从属关系的分句；例 68 的 brief 则译成描述性的方式状语短语。

当作定语的分词短语含有时间、条件、原因、让步等状语概念时，汉译时会把这样的分词短语译成状语从句。

例 69　Liquids，containing no free electrons，are poor conductors of heat.

【译文】各种液体，由于不含自由电子，所以是热的不良导体。

例 69 中 containing no free electrons 在原句中作主语 liquids 的定语，与整个句子在意义上构成因果关系，故译为原因状语。

例 70　A coil of wire moving in a magnetic field will have an e. m. f. induced in it.

【译文】当线圈在磁场中运动时，内部会感应起一个电动势。

例 70 中的 a coil of wore moving in a magnetic field 是 have an e. m. f. induced in it 的实现条件，故可以把分词短语转译为条件状语"当线圈在磁场中运动时"。

6.2.1.6　状语的译法

在英语中，状语用来修饰动词、形容词和副词，副词、介词、分词短语、不定式短语等都可以作状语，其主要译法如下：

（1）直译

原句中的状语译成汉语句的状语，可以视情况在语序上进行调整。

例 71　The charge passes at the uniform rate.

【译文】电荷匀速通过。

例 72　They have already fulfilled the electric energy production targets.

【译文】他们已经完成了发电量。

例 71 中的 at the uniform rate 译成方式状语，而例 72 中的副词 already 译成程度状语。

例 73 China's hydro power and coal resources are not evenly distributed.

【译文】中国的水力资源和煤炭资源分布不均衡。

例 73 中的 evenly distributed 译成主谓短语"分布不均衡","不均衡"是"分布"的方式状语。

(2) 转译为汉语的主语

下面的句子是科技英语中常见的固定句型,其翻译模式也是很固定的。

例 74 The device is shown schematically in Fig. 8.

【译文】这种装置的简图如图 8 所示。

例 75 The blueprint must be dimensionally correct.

【译文】蓝图的尺寸必须正确。

例 74、例 75 中的状语 schematically 和 dimensionally 分别由名词 scheme 和 dimension 派生而来,汉译时将词性还原。

例 76 The digital computer, in its computation section, can do mainly two things: add and subtract.

【译文】数字计算机的计算装置主要能做两件事——加和减。

例 77 It is stated in Ohms Law that the current flowing in a circuit is equal to the applied voltage divided by the resistance.

【译文】欧姆定律指出,电路中的电流等于外施电压除以电阻。

例 78 A nuclear power plant will be constructed in the region.

【译文】该地区将建一个核电厂。

英语中许多作状语的介词短语,在汉译时可译为句子的主语,此时介词一般省译。例 76、例 78 中的状语都由"介词 + 名词短语"构成,翻译时将介词省略,介词后面的宾语译作句子的主语。

(3) 转译为汉语的定语

英语中的一些介词短语(其中最常见的是由 in 引导的)在意义上与句子中的某个名词关系密切,二者存在着修饰与被修饰的关系。汉译时,这样的介词短语常常可译成该名词的定语。

例 79 The temperature coefficient of the resistance is positive for metals and negative for semiconductors.

【译文】金属的电阻温度系数是正的,而半导体的是负的。

例 79 中的介词短语 for metals 和 for semiconductors 与主语 the temperature coefficient of the resistance 关系密切,因此,汉译时这两个介词短语分别译为定语"金属的"、"半导体的"。

例 80 In Britain, the first stand-by gas-turbine electricity generator was in operation in Manchester in 1952.

【译文】英国的第一台辅助燃气轮发电机于 1952 年在曼彻斯特开始运行。

例 80 中的介词短语 in Britain 与主语 the first stand-by gas-turbine electricity generator 关系密切,因此,汉译时该介词短语译作定语"英国的"修饰"第一台辅助燃气轮发电机"。

例 81 The typewriter is chiefly characterized by its portability.

【译文】这部打字机的主要特点是携带方便。

例81中的状语 chiefly 修饰谓语 characterized，随着谓语转译成主语“特点”，状语也相应的转译为修饰主语的定语“主要”。

6.2.2 并列句的译法

英语的并列句主要由并列连词 and、but、or 等把两个或两个以上简单句连接起来。

6.2.2.1 直译

大部分并列句都可以采用直译法来翻译，并列连词可以视具体情况译出或省译。

例82 We could use two resistors in series, or we could increase the value of the present single resistor.

【译文】我们可以用两个串联的电阻，或者我们可增加现有的一个电阻的电阻值。

例83 Insulators in reality conduct electricity but their resistance is very high.

【译文】绝缘体实际上也导电，但它们的电阻很大。

例84 An alternating current has not a constant direction, and it has no constant magnitude.

【译文】交流电没有固定的方向，并且大小也在变化。

例85 The turns of line must be insulated from each other and from the iron, or the current will flow through short-circuits instead of flowing around the coil.

【译文】线圈的各线匝之间以及与铁心之间均应绝缘，否则，将产生短路而不能形成流经线圈的电流。

上述例子中的连词基本上都是按照字面意思直译，这也是科技英语中最直接最简单的方法。

6.2.2.2 译成复合句

英语并列句译为复合句主要有以下几种情况：

（1）若 and 连接的前一个分句是祈使句，这个祈使句表示条件的含义，可以译成条件状语从句，并视具体条件增译“如果”、“要使”、“假若”等。

例86 Close an electric circuit and electricity begins flowing.

【译文】如果闭合电路，电流就开始流动。

例87 Use a transformer and power and low voltage can be changed into power at high voltage.

【译文】如果使用变压器，低电压就可能转变为高电压。

以上两个例子中，and 连接的两个句子具有条件关系，译成“如果……就……”。

（2）当 and 表示结果或某种符合逻辑的过程时，转译为“从而”、“因而”等。

例88 The amount of power being used changes during the day, and the job of the control engineer is to switch power stations in and out.

【译文】一天内用电量是变化的，因而调度人员的工作就是让一些电厂投入或退出运行。

例89 The number of protons is equal to that of electrons and the whole atom is electrically neutral.

【译文】质子数与电子数相等，因而整个原子呈电中性。

从语义层面和逻辑关系上看，例 88、例 89 中的前一个分句是后一个分句的原因，故两个分句有一定的因果关系，中间的连词 and 译成“因而”。

（3）当 and 表示转折或对比时，转译为“但是”、“于是”、“然而”等，语气比 but 弱。

例 90 He is skeptical about the veracity of the data, and to me it seems no problem.

【译文】他对该数据的真实性表示怀疑，不过在我看来没有什么问题。

（4）当 and 表示一种伴随关系，或作进一步说明时，转译为“而且”。

例 91 Storage batteries are not effective for prolonged use, and they are expensive.

【译文】蓄电池不能长期有效地使用，而且很昂贵。

例 91 中，由于 and 前后所表述的两个特点是蓄电池都具备的，不是或缺的关系，所以可以用“而且”表示递进的逻辑关系。

（5）for 作为并列连词时常常译为“因为”，但语气不强，只是对前一个分句做出补充性说明，或表示一种推测的理由。

例 92 There exist neither perfect insulators nor perfect conductors, for all substances offer opposition to the flow of electric current.

【译文】没有绝对的绝缘体也没有绝对的导体，因为一切物体对电流都有阻力。

例 92 中，由 for 引导的从句在汉译时被译成了一个表示补充说明意思的分句。

6.3 名词性从句的翻译

英语中的名词性从句主要指主语从句、宾语从句、同位语从句和表语从句，这些从句在整个句子中的功能相当于名词或名词性短语，其引导词主要包括三种：连接代词（who、whoever、whom、whose、what、whatever、which、whichever）；连接副词（when、where、why、how）；从属连词（that、whether、if）。

6.3.1 主语从句的译法

主语从句是在主句中充当主语的句子，从结构上看，主要有两种形式：“主语从句 + 谓语 + 其他成分”和“it（形式主语） + 谓语 + that 引导的主语从句”。

6.3.1.1 “主语从句 + 谓语 + 其他成分”结构的译法

由关联词 what、who、whenever、wherever、whatever、however 等和从属连词 that、whether 等引导的主语从句，一般采用顺译法，即将英语的主语从句译成汉语的主语，并置于句首，与英文的原文顺序基本一致。

例 93 What we are talking about is how to make a communication between the MicroVax 3800 and the plant's DCS through the scada system.

【译文】我们正在讨论的是如何通过管理控制和数据采集系统（scada）在 MicroVax 3800 和装置的 DCS 系统间建立通信。

例 94 That substances expand when heated and contract when cooled is a common physical phenomenon.

【译文】物体热涨冷缩是普遍的物理现象。

以上两例中的主语从句是为了给主语更多的强调，同时也为了使句子前后平衡。由于句中的主语从句不长，所以直接译成了汉语句子中的主语。

例 95　How energy and matter behave, how they interact one with the other, and how we control them to serve the people make up the basis of two physical sciences—physics and chemistry.

【译文】能和物质如何起作用，它们如何相互影响，我们如何对其控制以使之为人类服务，这些问题构成了两门基础自然科学—物理学和化学的内容。

例 95 中的主语从句由几个从句并列，在翻译时先叙述主语从句的内容，之后再用“这些问题”做同位语重新提及一下。

6.3.1.2　“it（形式主语）+谓语+that 引导的主语从句”结构的译法

在英语中，更多的是采用 it 作形式主语，而用 that 引出的主语从句作真正的主语。这种结构的句子，翻译时视情况可以提前，也可以不提前。

（1）照译

例 96　It is a fact that electronic computers may be used as process control units.

【译文】事实上，电子计算机可用作工艺过程控制装置。

例 97　It is important that science and technology be pushed forward as quickly as possible.

【译文】重要的是尽快把科学技术搞上去。

例 98　It is obvious that oil is lighter than water.

【译文】显然，油比水轻。

例 99　It is evident that a well lubricated bearing turns more easily than a dry one.

【译文】显然，润滑好的轴承，比不润滑的轴承容易转动。

例 100　It seems that these two branches of science are mutually dependent and interacting.

【译文】看来这两个科学分支是相互依存，相互作用的。

例 101　It is possible that AC can be transformed into DC by using a rectifier.

【译文】可以用整流器将交流电转换成直流电。

以上例子中，主语从句置于主句之后，形成了主句在前、从句在后的格局，翻译时一般都采用顺译法，省译形式主语 it，句子译成无人称句。

（2）倒译

倒译，即先把从句译为主语，而后将原来主句中的谓语部分照译为谓语。为了强调，it 一般可以译出来。

例 102　It is still to be discussed where the new substation will be built.

【译文】新变电站建在什么地方（这一问题）仍有待讨论。

例 103　It is certain that we shall produce this kind of engine.

【译文】我们将生产这种发动机，这是肯定无疑的。

例 104　It is a general rule that gases, like liquids, always flow from regions of high pressure to regions of lower pressure.

【译文】气体像液体一样，总是从高压区流向低压区，这是一个普遍的规律。

以上例子中，都将后置的主语从句前置译出且在译文中作主语。同时，为了表示强调，作形式主语的 it 也需译出。

(3) 合译

当主句为"It + vi."、"It is + adj."时，主句可缩译为状语，与从句一起合译为简单句。

例 105 It appears that this is the only exception to the rule.

【译文】这似乎是这条规则的惟一例外。

例 106 It is apparent that the design engineer is a vital factor in the manufacturing process.

【译文】设计工程师显然在制造过程中起着主要的作用。

以上例中的主语从句被译成主句，而主句 it appears 和 it is apparent 分别缩译为副词"似乎"和"显然"作状语，与从句重新组合成一个简单句。

6.3.2 宾语从句的译法

英语宾语从句有四种类型，即动词宾语从句、介词宾语从句、用 it 作形式宾语的宾语从句和用作直接引语的宾语从句。具体译法如下：

6.3.2.1 顺译

例 107 How do you know if a substance contains acids?

【译文】你怎么知道一种物质是否含有酸呢?

例 108 Scientists have proved it to be true that the heat we get from coal and oil comes originally from the sun.

【译文】科学家已证实，我们从煤和石油中得到的热都来源于太阳。

6.3.2.2 倒译

介词 except、besides 和 but 的宾语从句，在英语中通常都放在句末。按照汉语习惯一般应采用倒译法，将其译文安排在主句之前，有时在译文中也可将 that 引起的宾语从句提前。

例 109 Engineering metals are used in industry in the form of alloys except that aluminum may be used in the form of a simple metal.

【译文】除了铝可以纯金属形态大量使用之外，各种工程金属都是以合金形式应用于工业。

例 109 中 that 引导的宾语从句作介词 except 的宾语。需要注意的是，位于介词 besides、except、but 后的宾语从句，汉译时通常需要前置，可以译作并列分句，用"除了……"、"除……之外"、"只是……"、"但是……"等来引导。

例 110 Of all the methods, electrostatic precipitation is unique in that the electric forces are large.

【译文】在所有的方法中，只有静电集凝法由于静电作用力大而与众不同。

例 110 中的 in that 一般被用作一个表示因果关系的连词，译成引导原因状语从句的连词"由于……"。

6.3.2.3 转译

介词宾语从句有时可据其逻辑含义转译为原因状语从句或并列从句。

例 111 The molecules of a liquid are much different from grains of sand in that they attract one another with considerable force.

【译文】液体的分子和沙子的颗粒有很大的不同，这是因为液体的各个分子以很大的力相互吸引。

例112　Liquids are different from solids in that liquids have no definite shape.

【译文】液体与固体不同，这是因为液体并没有一定的形状。

如果在“因为”前面添加“这是”两字，则构成并列复合句。

6.3.3　表语从句的译法

表语从句是位于主句的连系动词后面，由从属连词 that、whether、because 和带疑问意义的关联词 who、what、which、how、when、where、why 等引导的从句，其译法比较简单，一般按照原文的顺序翻译成汉语的判断句。

6.3.3.1　顺译

例113　An important character of radiation is that it can occur in a vacuum.

【译文】辐射的一个重要特点是它能够出现在真空里。

例114　The low resistance in pure copper wires is why they are so widely used in electric circuits.

【译文】纯铜导线的电阻低，这就是它在电路中广泛应用的原因。

以上例子中，be 后面表语从句的中心词都是名词，作判断动词“是”的宾语，整个句子的结构是判断句。

6.3.3.2　例译

在 This is why... 和 This is what... 句型中，如果表语从句太长，从汉语修辞考虑，也可将从句倒译在句子前半部，将主句译在后面，分别译为“原因（理由）就在这里”和“就是这个意思（道理）”等。

例115　That is why practice is the criterion of truth and why “the stand point of life, of practice, should be first and fundamental in the theory of knowledge”.

【译文】所谓实践是检验真理的标准，所谓“生活、实践的观点，应该是认识论的首先的和基本的观点”，理由就在于此。

6.3.4　同位语从句的译法

同位语从句在科技英语中比较常见，同位语从句用来对它前面的名词（或代词）作进一步的解释，由从属连词 that、whether 引导的从句，其常见的先行词为 idea、evidence、fact、thought、theory、news、conclusion、problem、opinion、statement、belief、rumor、suggestion、order、decision、certainty 等。在实际翻译中，主要译法如下：

6.3.4.1　译成汉语中独立的后续分句

例116　These uses are based on the fact that silicon is a semi-conductor.

【译文】这些用途是基于一个事实，即硅是半导体。

例117　And there was possibility that a small electrical spark might accidentally bypass the most carefully planned circuit.

【译文】而且总有这种可能性：一个小小的电火花，可能会意外地绕过最为精心设计的线路。

例 118 The problem whether the quality of the products is up to requirement has not been settled.

【译文】有个问题尚未解决，即产品质量是否合乎要求。

例 119 We are familiar with the idea that all matter consists of atoms.

【译文】我们都熟悉这样一个概念，即一切物质都是由原子组成的。

例 120 At the end of the last century, an important discovery was made, that everything was partly built of electrons.

【译文】上世纪有一个重要发现，即一切事物都有一部分是由电子构成。

以上例子中 that 引导的同位语从句译成了一个独立分句，采用“即……”、“就是……”的形式，或用破折号或冒号将其引出，将同位语从句顺译在先行词的后面。该方法尤其适用于较长的同位语从句的翻译。

6.3.4.2 转译成汉语的其他形式的句子成分

例 121 Even the most precisely conducted experiments offer no hope that the results can be obtained without any error.

【译文】即使进行的是最精确的实验，也没有希望获得无任何误差的实验结果。

例 121 中的同位语从句转译成让步状语从句的后续分句。

例 122 The thought suddenly came to the inventor's mind that the problem would be moderated to some degree if he added a few more components to the device.

【译文】发明家突然想到，如果在装置上再加上几个组件，问题就可得到某种程度的缓解。

例 122 中的同位语从句译成宾语分句。这里需要指出的是：当英文句子中同位语从句的先行词是 belief、assumption、suggestion、hope 等具有动词意义的名词时，一般将这些名词转译为动词。相应地，这类名词后面的同位语从句被转译成汉语的主谓短语，相当于英语中的宾语从句。

例 123 The fact that electric currents passing through a wire produce magnetic field is known to all.

【译文】电流通过导线产生磁场这一事实是众所周知的。

例 124 The chance that a circuit breaker may fail to break up should not be higher than that for protection and trip.

【译文】断路器失灵的机率不应高于保护和跳闸回路失灵的机率。

以上两例均把同位语从句转译成了定语。因为同位语从句不是定语从句，但在意义和形式上都接近于定语从句。

6.4 定语从句的翻译

在科技英语中，表达一个概念，讲述一个道理或现象时都必须准确无误，所以常使用定语从句。英语中的定语从句不仅结构复杂，而且含义繁多，具有补充、转折、原因、结果、目的、条件、让步等意义。翻译时，根据定语从句的不同结构和不同含义，可采用不同的译法。

6.4.1 语法层次

从主句与从句的关系来看，科技英语的定语从句大体上分为限定性定语从句（restrictive attributive clause）和非限定性定语从句（non-restrictive attributive clause）。从语法层次来看，定语从句在句中的语法功能是修饰、限制先行词。因此，在翻译时，关键在于先找出先行词，再采用前置法、后置法、溶合法等方法翻译。

6.4.1.1 前置法

前置法是指把英语定语从句译成带“的”的定语词组，大都适用于限定性定语从句。把英语限定性定语从句译成带“的”的定语词组，放在修饰词之前，从而将复合句译成汉语单句。一些较短而具有描写性的英语非限定性定语从句，也可译成带“的”的前置定语，放在被修饰词前面。如：

例 125 The reason why things don't fall off the earth is rather simple.

【译文】物体不掉离地球的原因是相当简单的。

例 126 Language is a tool by means of which people communicate ideas with each other.

【译文】语言是人们赖以交流思想的工具。

例 127 A planetary gear is one whose teeth are cut out on the inside of a wheel instead of the outside.

【译文】行星轮是轮齿刻在轮子里边而不是外延的一种齿轮。

例 128 Yet, there exist complex computations in science and engineering which people are unable to make.

【译文】到目前为止，在科学和工程方面还存在许多人们无能为力的复杂计算。

以上例子中，定语从句均译成前置的形式：“……的……”。将定语从句译在所修饰的名词前面，这是一种比较简单的译法。

6.4.1.2 后置法

定语从句除了表示定语的意义外，有时还可以相当于并列分句。这种从句起着对主句的内容进一步阐述、说明的作用。英译汉时，一般将整个从句译成一个并列句，这种译法一般用于非限定性定语从句、以及某些结构太长的限定性定语从句。

（1）译成并列分句，重复英语先行词

例 129 Day light comes from the sun, which is a mass of hot, glowing gas.

【译文】日光来自太阳，太阳是一团炽热的发光的气体。

例 130 The concept of energy leads to the principle of the conservation of energy, which unifies a wide range of phenomena in the physical science.

【译文】能量的概念导致了能量守恒定律，该定律统一了物理科学中许多广泛的现象。

例 131 Another kind of rectifier consists of a large pear-shaped glass bulb from which all the air has been removed.

【译文】另一种整流器由一个大的梨形玻璃泡构成，泡内的空气已全部抽出。

以上例子中的非限定性定语从句分别都译为与前面主句并列的分句，并且由于表达的需要，在后面的分句中重复前面句子中的相关名词“太阳”、“该定律”和“泡”。

（2）将定语从句译为另一个分句，将关系代词译成人称代词“它”或“它们”，作该

分句的主语

例 132 If you deal with the younger age group, then you will see a lot of the acute infections such as herpes and trench mouth, which is due to bacteria and causes open sores between the teeth.

【译文】如果你给较小年龄组的人做检查，你就会遇见很多诸如疱疹和战壕口炎（坏死性溃疡性龈炎）等急性感染的病症，这些病症都是由细菌感染所引起的，并会导致牙齿之间的开放性溃疡。

例 133 These waves, which are commonly called radio waves, travel with the velocity of light.

【译文】这些电波以光的速度传播，它们通常被称为无线电波。

例 134 ISDN is the name given to a network that is able to transmit and switch a wide variety of telecommunication services.

【译文】综合业务数字网（ISDN）是一个网络系统，它传输和交换各种电信业务。

以上例子中，几个复合句的叙述层次和汉语结构相近，句中由 which 引导的非限定性定语从句对前面的先行词解释说明，在语序上采用顺译，译为第二个分句，译文中都采用代词来指代前面分句中的相关叙述。

（3）译成并列分句，省略英语先行词

例 135 Sulfur melts at a temperature of 112. 8 °C, where it changes to yellow liquid.

【译文】硫在 112. 8 °C，变成黄色的液体。

例 136 The last big Alaskan earthquake created a tsunami, which could be felt 1,500 miles away.

【译文】最近发生的阿拉斯加大地震引起了海啸，在 1,500 英里之外都能感觉到。

例 137 The electricity is changed into the radio-frequency power which is then sent out in form of radio waves.

【译文】电转变成射频能，然后以无线电波的形式发射出去。

以上例子中的非限定性定语从句译成了主句的并列分句，根据表达通顺、简练的需要，分句中省译了前面句子中的先行词。

6. 4. 1. 3 溶合法

溶合法是把原句中的主语与定语从句溶合在一起译成一个独立句子的一种翻译方法。

（1）将主句和从句译成一个句子

这种译法一般是把主句的主语作为译文的主语，从句作为译文的谓语。该译法特别适用于“there be + 先行词 + 定语从句”句型的翻译。将英语中这种句型转化成汉语句型的过程事实上就是一个将先行词和定语从句译成一个句子的过程。

例 138 There are some materials which possess the power to conduct heat.

【译文】某些材料具有导热的能力。

例 139 In a conductor there are a large number of electrons that move freely from atom to atom.

【译文】导体中有大量的电子在原子之间自由运动。

例 138 的主句 there are some materials 被缩译成一个短语作为主语，原定语从句的谓语译成整个句子的谓语。例 139 的主句 there are a large number of electrons 与定语从句融合在一起，译成一个独立句。

（2）将主句转译成名词短语作主语，定语从句译成谓语

例 140　OSHA has qualified the noise level in industry which has become a major concern for many digital controls manufacturers.

【译文】职业安全保健管理局规定的噪音水平已成为许多数字控制器制造商关注的主要问题。

例 141　The factory produced machine tools to which precision instruments were attached.

【译文】该厂家生产的机床附有一些精密仪器。

例 142　Good clocks have pendulums, which are automatically compensated for temperature changes.

【译文】好钟的钟摆可以自动补偿温度变化。

例 143　Gases, the molecules of which are widely separated from one another, have great compressibility than liquids.

【译文】气体的分子互相之间距离很远，因而气体的可压缩性较液体大。

例 140～142 中的定语从句不宜译成定语，因为句义重点在从句上。为了突出定语从句的内容，可以把从句译成谓语（包括与谓语有关的部分），而把主句压缩成汉语词组做主语。主句 OSHA has qualified the noise level in Industry、the factory produced machine tools 和 good clocks have pendulums，分别译成主语词组“这间厂生产的机床”和“职业安全保健管理局规定的噪音水平”，定语从句 which has become a major concern and consideration for many hydraulic controls manufacturers 和 to which precision instruments were attached 分别译成动词短语“已成为许多液压控制器制造厂商关注与考虑的主要问题”和“附有一些精密仪器”作谓语。例 143 中的定语从句在翻译时也译成了主谓短语作谓语。

6.4.2　语义层次

不管是限定性定语从句，还是非限定性定语从句，其中有些形式上是定语从句，但从语义逻辑关系上分析则起着状语从句的作用。在翻译这类从句时，译者要根据定语从句和先行词以及整个主从句的内在关系，将其译成相应的状语从句，表示原因、结果、目的、让步等。

6.4.2.1　转译成原因状语分句

例 144　To make an atom bomb we have to use Uranium 235, in which all the atoms are available for fission.

【译文】制造原子弹必须用铀 235，因为它的所有原子都可裂变。

例 145　We can study the motion of the projectile by watching the motion of its center of gravity, at which the mass of the projectile is considered to be concentrated.

【译文】我们可以通过观察弹丸重心的移动来研究弹丸的运动，因为弹丸的质量通常被认为是集中在重心上。

例 146　It (reliability) can be also improved by design for ease of maintenance, which involves ergonomic man/machine relationships affecting accessibility, the nature of adjustments, facilities for replacement, measurement and inspection, and by logical faultfinding procedures.

【译文】通过逻辑查错程序和便于维护的设计也能增加可靠性，因为这种设计需要考虑：人类工程学的人机关系，这些关系会影响维护的方便性、调整的性质、更换部件的便利及测量和检验。

例 147 A cat, whose eyes can take in many more rays of light than our eyes, can see clearly in the night.

【译文】由于猫眼比人眼能吸收更多光线，所以猫在夜里也能看得很清楚。

例 148 A solid fuel, like coal or wood, can only burn at the surface, where it comes into contact with the air.

【译文】固体燃料，如煤和木材，只能在表面燃烧，因为表面接触空气。

以上例子中的定语从句从语义层次上看，分别给出了主句"制造原子弹必须用铀235"、"通过观察弹丸重心的移动来研究弹丸的运动"、"过逻辑查错程序和便于维护的设计也能增加可靠性"、"猫在夜里也能看得很清楚"和"固体燃料只能在表面燃烧"的理由，因此可以转换成原因状语从句。特别是例 146 中 by design for ease of maintenance 与 by logical faultfinding procedures 是并列的关系，但在 by design 后面跟有很长的定语从句。如果也按这种顺序组织译文，则会模糊并列关系，不利于表达原文的意义。故将定语从句翻译成状语从句，并转移到并列成分后，同时，将定语从句中 affecting 引导的现在分词短语转换成分句，将并列成分位置前后颠倒，使译文中重复的词紧紧相连。

6.4.2.2 转译成结果状语分句

例 149 A liquid heated in vessel expands relatively to the vessel, which also expands.

【译文】在容器中加热的液体会相对于容器而膨胀，结果使容器也膨胀。

例 150 If clouds are cooled further, their droplets of water form big drops that fall as rain.

【译文】如果云块继续受冷，云中的小水滴就会形成大水滴，最后变成雨落下来。

以上例子中的定语从句主要补充说明主句所叙述内容的结果，故可以转译成结果状语。

6.4.2.3 转译成让步状语分句

例 151 A gas occupies all of any container in which it is placed.

【译文】气体不管装在什么容器里，都会把容器充满。

例 152 Electronic computers, which have many advantages, cannot carry out creative work and replace man.

【译文】尽管电子计算机有许多优点，但它们不能进行创造性工作，也不能代替人。

根据原句中的定语从句与主句在语义上的逻辑关系，翻译时将两个例句中的定语从句译成让步状语分句。

6.4.2.4 转译成目的状语分句

例 153 Following the side-band filter is an amplifier which makes up for the filter insertion loss.

【译文】边带滤波器后面接了一个放大器，用来补偿这边带滤波器的插入损耗。

例 154 Ultrasonic waves produce pulsed signals, by means of which various defects in metal can be detected.

【译文】超声波能产生脉冲信号，用以检测金属中各种缺陷。

例155 An improved design of such a large tower must be achieved which results in more uniformed temperature distribution in it.

【译文】这种大型塔的设计必须改进，以使塔内温度分布更均匀。

以上例子中的定语从句从语义层面上看，是主语行为或动作状态的目的或意图，故翻译时转译为目的状语从句。

6.4.2.5 转译为条件分句

例156 A body that contains only atoms with the same general properties is called an element.

【译文】物质如果包含的原子性质都相同，则称之为元素。

很显然，例156中定语从句表达的含义"物质包含的原子性质都相同"是"称之为元素"的条件，故定语从句在翻译时转成条件分句。

6.4.3 as 引导的限制性定语从句的译法

关系代词 as 经常与 such、the same、as many、as much 等配合使用。在这些句型中，as 引导限制性定语从句，在从句中作主语、表语、宾语等，以 as 引导的这类限制定语从句往往有比较固定的一些译法。

6.4.3.1 such +（名词）+as 或 such as 的译法

例157 Such liquid fuel rockets as are now being used for space research have to carry their own supply of oxygen.

【译文】像现在用于宇宙研究的这类液态燃料火箭，必须自带氧气。

例158 Such propellers as we have recently designed for small ships are actually modeled on fish tails.

【译文】像我们近来为小船设计的那样的螺旋桨实际是模仿鱼尾制造的。

such +（名词）+as 或 such as 译为"像……之类的"、"像……（这）那样的"、"……的一种"等。

6.4.3.2 the same... as 的译法

例159 Many inventors followed the same principles as that French inventor had used in his invention.

【译文】许多发明家遵循那个法国发明家在发明中曾用过的同样的原理进行发明创造。

例160 A color transmission contains the same information as a black and white transmission.

【译文】彩色传输所容纳的信息，和黑白传输容纳的信息一样。

6.5 状语从句的翻译

一般来说，英语状语从句与汉语偏正复合句相对应，但是偏正复合句中的分句不含时间、地点、比较等意义，另外绝大多数连接时间、地点、比较状语从句的关联词在汉语中为动词、介词或者副词，而不是关联词语，不能用来连接句子。因此，这些状语从句一般不能与汉语句子相对应，而相当汉语单句中由词组充当的成分。但有时候，地点、时间状语从句与汉语某些偏正复句中的单句相当；某些关联词连接的比较状语从句又与汉语联合复句中的分句或者偏正复句中的分句相当。表示原因、结果、目的、让步、条件等意义的状语从句与汉语偏正复句中的分句或者正句相对应，但是也不尽相同，如目的、结果状语从句有时候又与汉语单句成分相对应。

6.5.1 时间状语从句的译法

用来引导时间状语从句的连词有 when、as、while、before、after、since、until、by the time、each time、every time、as soon as、instantly、immediately、directly、the moment、once、no sooner... than...、scarcely/hardly... when...、whenever 等。时间状语从句的翻译，主要在于掌握好各种时间意义的连词，正确理解和区别连词的含义。另外，按照汉语的习惯，时间 状语从句要放在其主句的前面，不管原文是自然语序还是倒装语序。但有的时候也可将主句后面的地点状语从句倒装译在主句的前面。

6.5.1.1 直译

时间状语从句一般与汉语单句成分相当。连接这些从句的关联词，如 when（当……时候）、before（在……之前）、where（在……地方）等在汉语中为“介词 + 宾语 + 名词”，构成介宾词组，也可构成偏正词组，如，“……时”（when）、“……之后”（after），在句中作状语。

例 161 Electric charges work when they move.

【译文】当电荷运动时，就做功。

例 162 As the sphere becomes larger, the waves become weaker.

【译文】随着范围的扩大，电波变得愈弱。

例 163 Current stops flowing as soon as we open the circuit.

【译文】电路一被中断，电流就停止流动。

6.5.1.2 转译

从主、从句之间的逻辑意义上讲，时间状语从句有时含有条件、假设、因果等意义。汉译时，可转换为汉语条件、假设或者因果等偏正复句中的分句，有时候还可以译成汉语单句中的词组。

例 164 Turn off the switch when anything goes wrong with the machine.

【译文】如果机器发生故障，就把开关关掉。

例 165 A body at rest will not move until a force is exerted on it.

【译文】若无外力的作用，静止的物体不会移动。

以上例子中的时间状语从句都转换为汉语中的条件状语从句。

例 166 When atoms split, the process is called fission.

【译文】原子分裂，其过程称之为裂变。

例 166 中的时间状语从句译成了与主句的并列分句。

6.5.2 地点状语从句的译法

引导地点状语从句的多为 where 和 wherever。当 where 引导的从句位于句首时，是一种加强语气的说法，而且含有条件意味，可译成“哪里……哪里……”。wherever 引导的从句位于句首时，除了有强调的意味外，还有“让步”的意义，可翻译成“不论到哪里；哪里都……”。

6.5.2.1　直译

例167　Where there are magnetic forces，there are poles.

【译文】哪里有磁力，哪里就有极。（或：有磁力的地方就有极。）

例168　Wherever conductors are needed，insulators will be indispensable.

【译文】凡是需要导体的地方，绝缘体就必不可少。

6.5.2.2　转译

（1）转译成短语

例169　As noted previously，where high reactivity of the charge makes close temperature control mandatory，the catalyst in the first reactor can be divided into two beds，with an intermediate liquid quench.

【译文】前面已指出，在装料有很高反应能力而必须严格控制温度的场合，第一反应器中的催化剂可分为两层，并在中间采用液体冷却。

例170　We use insulators to prevent electrical charges from going where they are not wanted.

【译文】我们使用绝缘子是为了防止电荷跑到不需要的地方去。

例169的地点状语从句转换为汉语中的介词词组“在……场合”，在译文中作地点状语。例170的地点状语从句译成汉语时转换为宾语。

（2）转译成条件状语分句

例171　Where the volt is too large a unit，we use the millivolt or microvolt.

【译文】如果用伏特作单位太大，我们可用毫伏或微伏。

例172　Cracks will come out clean when treated by ultrasonic waves.

【译文】如果以超声波处理，缝隙就会变得很洁净。

（3）转译成汉语的并列分句

例173　After the successful pressure test of pipe system，the insulation can be done up to the point where all flanges are still visible for later inspections and tightening-up under warm condition during the start-up phase.

【译文】在成功地进行了管道系统的压力试验后，能够将管道绝热层做到这样的程度，即所有凸缘仍然露在外面，以便以后于开工阶段在管道受热状态下加以检查和拧紧。

（4）转译成结果状语分句

例174　It is hoped that solar energy will find wide application wherever it becomes available.

【译文】太阳能将得到广泛的利用，以致任何地方都可以使用。

where或wherever引导的从句在逻辑上具有结果状语的意义，可转译成结果状语分句。

6.5.3　条件状语从句的译法

英语中连接条件状语从句的连接词有if（如果）、unless（除非，如果不）、providing that（假如）、so long as（只要）、on condition that（条件是）、suppose that（假如）、in case（如果）、only if（只要）、if only（但愿，要是... 就好了）等。

6.5.3.1 译成“先条件后结果”的句子结构

英语表达中，条件状语从句可以放在主句前或主句后，位置较随意。而按照汉语的习惯，不管表示的是条件还是假设，分句要放在复句之前。因此，英语的条件从句翻译成汉语时多遵循“先条件后结果”的表达模式。

例 175 If something has the ability to adjust itself to the environment, we say it has intelligence.

【译文】如果某物具有适应环境的能力，我们就说它具有智力。

例 176 Unless the air is first removed from a light bulb, the filament will burn up.

【译文】除非先把灯泡里的空气抽走，否则灯丝就会烧掉。

例 177 As long as light travels through one kind of substance, it proceeds in all direction in straight lines from its source.

【译文】只要光是在一种介质中传播，它就从光源笔直向各个方向照射。

例 178 The average speed of all molecules remains the same so long as the temperature is constant.

【译文】（只要）温度不变，所有分子的平均速度就不变。

例 179 If all the impeding forces of gravitation and resistance could be removed, there is no reason why the ball, once in motion, should ever stop.

【译文】如果能克服一切起阻碍作用的重力和阻力，就没有理由认为球一旦处于运动状态会再停下来。

例 180 A watch has to have gears and other moving parts as long as it has a dial with hands to tell the hours, minutes and seconds.

【译文】只要钟表有带指针的表盘来计时、分、秒，就必须有齿轮和其他运动部件。

例 181 If two lines are parallel, the corresponding angles are equal.

【译文】两直线平行，同位角相等。

以上例子中翻译时均按照汉语先条件后结果的表达模式翻译，将条件状语从句翻译在主句之前。其中，例 180 的译文中为了表达的简洁省略了关联词。

6.5.3.2 译成“先结果后条件”的结构

有时根据具体的语境中语义表达的重点不同，也可将条件状语分句放在译文中主句之后。

例 182 This kind of plane cannot be built unless we find a metal even lighter than this high-strength aluminum alloy.

【译文】我们不可能造出这种飞机来，除非我们找到一种比这种高强度的铝合金还要轻的金属。

例 183 Iron or steel parts will rust, if they are unprotected.

【译文】铁件或钢件是会生锈的，如果不加以保护的话。

以上例子中的条件从句表达的是一种补充说明情况的意义，汉译时可放在主句的后面。

6.5.3.3 条件虚拟语气的译法

例 184 Were there no transformers to adjust the voltage, long-distance transmission of electricity would be impossible.

【译文】若无变压器来调节电压，远距离输送电力是不可能的。

例 185 Should there be urgent situations, press this red button to switch off the electricity.

【译文】万一情况紧急，按这个红色按钮以切断电源。

英语的非真实条件句中，谓语用虚拟语气。这类从句一般放在句首，汉译时采用顺译法。当省略从属连词 if，而将 were、had、should 等移至主语前引起倒装语序时，汉译时将这类从句放在句子的前面。

6.5.4 比较状语从句的译法

引导比较状语从句的连词有 than（比，不，非）、as... as（和……一样）、not so (as)...as，the... the...、as if（好像，仿佛）等。

6.5.4.1 直译

连接比较状语从句的大多数关联词在汉语中不是关联词，不能连接句子，只能与其他词类搭配，构成词组，在句中充当某一成分。汉译时，可将比较状语从句转换为状语、定语或者谓语部分，关联词转换为动词、介词或者副词等。

例 186 The temperature at the sun's center is as high as 10,000,000 °C.

【译文】太阳中心的温度高达一千万摄氏度。

例 187 The outer portion of the wheel may travel as fast as 600 miles per hour.

【译文】轮子外缘的运动速度可能高达每小时 600 英里。

例 188 The oxygen atom is nearly 16 times heavier than the hydrogen atom.

【译文】氧原子的重量几乎是氢原子的 16 倍。

例 189 Mercury weighs more than water by about 14 times.

【译文】水银比水重约 14 倍。

以上例子中的方式状语按照字面意思译成谓语。

例 190 Solar cells are as different from so called solar heating panels as solid state physics is from plumbing.

【译文】太阳能电池与所谓的太阳能加热板的差别，正如固体物理学与波导设备的差别一样。

例 190 中的方式状语从句按照原句的语法结构翻译，对应译成方式状语从句。

6.5.4.2 结构转译

连接比较状语从句的关联词 rather than 的词义与汉语中连接复合句的关联词语“与其……不如……”、“而（不）”相同，因此，可将关联词 rather than 连接的比较状语从句转换为联合复句中的分句。

例 191 In such occasions we would rather increase the friction of the surface than decease it.

【译文】在这些情况下，与其减少，倒不如增加表面摩擦力。

在英文原句中，than 是并列连词，而其后的并列句却被翻译成了“与其……”这个表示次要的分句，译文的重点集中在“倒不如……”。

例 192 The thicker is the wire, the smaller is the resistance.

【译文】导线越粗，电阻越小。

“the + 比较级 . . . the + 比较级……”在英文中的结构是并列的，而在汉语译文中，前面的分句转译成了条件句，后面的分句是主句，原本两个并列关系的分句被转译成了汉语的主、从句。

6. 5. 5 原因状语从句的译法

英语中，原因状语从句的连接词有 because（因为）、since（既然，由于）、as（因为）、now that（既然）、seeing that（既然）、considering that（考虑到，因为）、in that（在某方面）、in view of the fact that（鉴于）、for the reason that（因为）等。这类状语从句翻译时一般放在主句之前，但是当原文的原因从句出现主句之后时，也可根据具体语境采取顺译法。

例 193 Since transistors are extremely small in size and require only small amount of energy, they can make previously large equipment much smaller.

【译文】由于晶体管的体积非常小，而且只需要少量的电能，它们能使庞大的设备体积缩小许多。

例 194 Because energy can be changed from one form into another, electricity can be changed into heat energy, mechanical energy, light energy, etc.

【译文】由于能量能从一种形式转换为另一种形式，所以电可以转变为热能、机械能、光能等。

例 195 The crops failed because the season was dry.

【译文】由于气候干燥，作物歉收。

例 196 A gas differs from solid in that it has no definite shape.

【译文】气体不同于固体是因为（就在于）它没有固定的形状。

例 197 The material first used was copper for the reason that it is easily obtained in its pure state.

【译文】最先使用的材料是铜，因为易于制取纯铜。

例 193 ~ 195 在翻译时都是按照汉语先因后果的表达习惯组织译文；而例 196 和例 197 则按照原句的顺序组织译文。

6. 5. 6 结果状语从句的译法

引导结果状语从句的连词有 so that、so. . . that、such that、such. . . that、to the extent that、to such an extent that、to such a degree、in as much as（因为，由于）、in so much as 等。由于这类从句在英语和汉语里都是放在主句之后，翻译时可采用顺译法。但应注意不能拘泥于引导结果状语从句的连词 so. . . that 等词义而一概翻译成“因而”、“结果”、“如此……以至于……”等，翻译时尽量避开连词，以免使译文过于欧化。

6. 5. 6. 1 直译

例 198 To reduce the waste of power, the iron core of a transformer is made of a large number of separate thin strips of iron coated with an insulating varnish, so that the eddy current cannot flow from one strip to another.

【译文】为了减少电力的浪费，变压器的铁心由许多涂有绝缘漆的单层薄铁片制成。因此，涡流不会从这一铁片流向另一铁片。

例 199 Electronic computers work so fast that they can solve a very difficult problem in a few seconds.

【译文】电子计算机工作速度如此迅速，一个很难的题目几秒钟之内就能解决。

例 200 Electricity is such an important energy that modern industry couldn't develop without it.

【译文】电是非常重要的一种能量，没有它，现代化工业就不能发展。

例 201 Some people cannot accept the idea that animals might have intelligence so that they are even more surprised at the suggestion that machine might.

【译文】有些人不能接受动物可能有智力的想法，因此，他们对机器可能有智力的设想就更加惊讶。

6.5.6.2 转译

例 202 The induced e.m.f. is in such a direction that it opposes the change of current.

【译文】感应电动势的方向是阻止电流发生变化的那个方向。

例 203 Light travels so fast that it is very difficult for us to imagine its speed.

【译文】光的传播速度快得使我们都难以想象。

例 204 The diameter of the pipe must be such that the liquid flows at a moderate speed.

【译文】管径的大小必须能使流体以适当的速度流过。

例 202 的译文中将结果状语从句转译为判断动词“是”的宾语分句；例 203 将结果状语从句译作使役动词短语作补语；例 204 将结果状语从句译作使役动词短语充当宾语。

6.5.7 目的状语从句的译法

引导目的状语从句的连词有 that（为了，以便）、so that（为了，以便）、lest（以防）、in case（以防，以免）、for fear that（以防）、in order that（为了）、in order that（为了）、for the purpose that（目的是）、in the hope that（以期）、to the end that（目的是）等。按照汉语的习惯，目的状语从句放在主句之前（“要想……”、“为了……”）或者之后（“以使……”、“以便……”、“以期……”）都可以，所以一般按照原文语序顺译。但有时候为了强调该从句所说的内容，在原文中目的状语从句位于主句之后时，也可采用倒译法，将从句译文放在主句译文之前。

例 205 We keep the battery in a dry place so that electricity may not leak away.

【译文】我们把电池放在干燥处，以免漏电。

例 206 Much work has been done in order that data can be transmitted by television and read off an ordinary TV screen.

【译文】已进行了大量的工作，使情报资料可以通过电视来播送并且在普通电视屏幕上显示出来。

例 207 Steel parts are usually covered with grease for fear that they should rust.

【译文】钢制零件通常涂上润滑脂，以防生锈。

以上例子中目的状语从句翻译时置于主句之后。英语的目的状语从句通常位于句末，汉译时可译成后置分句。

例 208 Usually the capacitor is made up of plates of large area so that large electrical charges may be stored.

【译文】为了能储存大量电荷，电容器通常用大面积的金属板来制造。

例 209 In order that the crops in open field could survive safely the winter frost prevention operation is of paramount importance.

【译文】为了使这些大田作物能够安全越冬，防霜作业具有极其重要意义。

以上例子中，为了强调目的，将目的状语从句置于句首。而汉语中表示“目的”的分句常用“为了”作为关联词置于句首，往往具有强调的含义。

6.5.8 让步状语从句的译法

在状语从句中，表示让步关系的连接词 though（虽然）、although（虽然）、even if/though/ when（即使）、as（尽管）、while（尽管）、no matter（不论，不管）、for all that（尽管）、granted that（即使）、in spite of the fact that（尽管）、despite the fact that（不管）、notwithstanding that（尽管）、whether... or（不管……还是……）、as（尽管）等。英语让步状语从句一般表示某种不利于主句某个动作的情况，但是并不影响主句的事实。汉语的转折偏正复合句中的分句表达一种意思，但是主句并不顺着前面分句的意思说下去，而是转向相反或者相对的方面。从表面看，这两种句子并不对应，但是从它们所用的关联词的含义来分析，可以看出这两种句子的意义相当，如 though/although（虽然……但是……）、while（尽管……但是……）、no matter how（不管怎样）、no matter what/whatever（无论）和 even if/though（即使）等。

6.5.8.1 直译

通常情况下，英语中的让步状语从句可翻译为汉语中的让步分句或无条件的条件分句。汉语中用以表示让步的关联词语有“无论……都/总是……”、“即使……也……”等，且汉语的让步分句多前置。

例 210 Even though robots can do many things than man does, they cannot replace man.

【译文】尽管机器人能做人所做的许多事情，但不能代替人。

例 211 Although technical advances in food production and processing will be needed to ensure food availability, meeting food needs will depend much more on equalizing economic power among the various segments of populations within the developing countries themselves.

【译文】尽管需要粮食生产和加工方面的技术进步来确保粮食的来源，满足粮食需求更多的也是取决于使发展中国家内部的人口各阶层具有同等的经济实力。

例 212 Iron and sulfur are the materials found most frequently in these deposits, although combinations of other elements may also exhibit pyrophoric properties.

【译文】尽管其他元素的组合也会阻止引火特性，铁和硫磺是这些堆积物中最常见的物质。

例 213 The chemical composition of water remains constant whether it is in solid, liquid or gaseous state.

【译文】无论处于固态、液态或气态，水的化学成份都保持不变。

例 214 All substances on the earth, whether gaseous, liquid or solid, are made up of atoms.

【译文】地球上所有的物质无论它们是气态的、液态的或是固态的，都是由原子组成的。

例 215 No matter what direction we choose in order to obtain information about the large-scale structure of the universe, we obtain the same answer as for any other direction.

【译文】为了获得宇宙的大尺度结构的信息，无论我们选择什么方向，我们所得到的答案对任何其他方向都是一样适用的。

例 216 However small it is, a body has weight.

【译文】一个物体无论多么小，都有重量。

例 217 The up-to-date computer can process readily however large a volume of data.

【译文】无论资料的数量如何惊人，现代计算机都能容易地加以处理。

以上例子中的让步状语从句翻译时都是按照“无论……都/总是……”、“即使……也……”和“尽管……也……”来组织译文的。

6.5.8.2 译为转折关系复句中的分句

例 218 The tsunami of 1957 killed no one in Hawaii, even though water levels were locally higher than 1946.

【译文】虽然 1957 年夏威夷海啸的局部水位比 1946 年还高，但是没有造成一人死亡。

例 219 Admitting that the vast expanse of oceans seems terrible, they provide us with abundant resources.

【译文】虽然浩瀚的大洋令人生畏，但它们向我们提供了丰富的资源。

例 220 Complicated as a modern machine is, it is essentially a combination of simple machines.

【译文】现代化的机器虽然复杂，但实质上不过是许多简单机器的组合。

例 221 All metals will melt though some require greater heat than others.

【译文】所有的金属都会熔化，虽然有的金属熔化时比其他金属需要更高的温度。

例 222 Weak as it is, the disturbance in the wind field is discernible.

【译文】该风场的扰动虽然很弱，但是还是可以看得出来。

例 223 Be ever bulky, the sun does not look larger than the moon from the earth.

【译文】虽然太阳体积如此之大，但从地球上看它并不比月球大。

例 224 Be that as it may, in practice these principles are likely to be interpreted so as to lead to the same ideas about the observed universe.

【译文】虽然如此，在实践中这些原理是可能解释的，从而使我们对观测的宇宙得出同样的认识。

以上例子中的译文都是采用“虽然……但是”的叙述模式。

就句子的次序来说，英语中的让步状语从句可前可后，而汉语中的转折、条件、假设偏正复句中的分句一般要前置，所以汉译时要调整句序。另外，英语中让步状语从句的关联词只能单独使用，但是汉语偏正复句的关联词一般成对使用。

6.5.9 方式状语从句的译法

引导方式状语从句的连词有 as、like、as if、as though、in a manner that、in this way that、in such a way that、to the extent that、to such an extent that、just as、according as、in degree as、in proportion as、in much the same way as 等。方式状语从句一般可按照原文顺序译为方式状语从句，有时候还可译为并列分句或者定语从句。

6.5.9.1 直译

例 225 Heat can flow from a hot body to a cooler one as if it were a fluid.

【译文】热能从一个热的物体传到一个较凉的物体上，好像流体一样。

例 226 When heated, gases expand as liquids and solids do.

【译文】气体受热时，像气体和固体一样，会发生膨胀。

例 227 The coil carrying currents has a magnetic field, as if it were a magnet.

【译文】载流线圈好像是磁铁一样具有磁场。

例 228 At first glance, this simplification might seem just as likely to result in underestimating a future benefit as it is in underestimating a future risk.

【译文】乍一看，这一简化似乎正如很可能导致低估未来的危险那样，导致低估未来的好处。

例 229 The result of this experiment is good enough as it is.

【译文】照现在这样，这个实验结果够不错了。

上述例子均采用直译法，原句中的方式状语在译文中仍体现为方式状语。

6.5.9.2 转译

例 230 Reading is to the mind what exercise is to the body.

【译文】读书之于精神，正如锻炼之于身体。

例 231 Each of two wires carrying currents has a magnetic field, as if it were magnet.

【译文】通电的两根导线每一根都有一个磁场，好像是一块磁铁。

例 230 含一个固定句型 A is to B what C is to D，意思是"A 之于 B，正如 C 之于 D"，虽然语法形式上是一个方式状语从句，但在语义层面上，是一个类比表达方式，两个分句呈并列关系。例 231 的方式状语从句译成了一个后续分句，与主句在译文中呈并列关系。

6.6 长句翻译的常用技巧

6.6.1 长句翻译的基本问题

大量使用长句是科技英语最重要的特征之一。理解和掌握英语长句的译法对于做好科技英语翻译工作非常重要，也是一个具有挑战性的难题。英语长句一般有如下几个特点：一是后置修饰语多；二是并列成分多；三是句法结构复杂、层次重叠。

由于英汉两种语言的差别，英语长句翻译涉及两个基本问题：英汉语序上的差异和英汉表达方式的差异。

6.6.1.1 英汉语序上的差异

英汉两个民族在思维上有差异，在语言表达方式上也有区别，主要表现为：汉语构句重逻辑顺序，属于自然语序，而英语构句比较灵活，自然语序与特异语序兼之；汉语的修饰词一般都是前置式，而英语中存在着大量起修饰作用的后置定语和后置状语，由于这些差异，当把一个含有多个后置修饰语的英语长句译成汉语时，就需按照汉语的语法和表达习惯安排、调整语序，尤其是定语和状语的语序。

例232 The single-line diagram summarizes the relevant information about the system for the particular problem studied.

【译文】单线图概括了所研究具体问题的有关系统的相关信息。

例句中介词短语和过去分词都作后置定语，翻译成汉语时应将其放在所修饰名词之前。

例233 Varies the generator terminal voltage in accordance with its load to hold constant voltage at some point electrically remote from the generator.

【译文】由于按发电机的负荷改变发电机的端电压，使远离发电机的某一电气点的电压稳定不变。

英语中某些单个形容词作定语时一般前置，但构成形容词短语时要后置，汉译时则要译成前置定语。另外，这个句子是一个倒装句，主语是 generator terminal voltage，谓语是varies。

6.6.1.2 英汉表达方式上的差异

除了在语序上的差异，英汉在逻辑表达上也存在着很大的差异。具体说，就是在行文层次和主次安排上。因此，在汉译时，需要按照汉语的逻辑思维方式来安排行文层次和主次关系。

(1) 有关时间先后关系的表述

例234 Hardly had the operator pressed the button when all the electric machines began to work.

【译文】操作员一按电钮，所有电机就开始运行。

例235 Switch it on after you' ve connected the electric kettle to an electrical circuit.

【译文】当把有水的电壶接入电路后，再打开开关。

上述的英语复合句中，表示时间的从句可以放在主句之前，也可以放在主句之后，汉语译文中则通常先叙述先发生的事，后叙述后发生的事。

(2) 因果关系的表述

例236 Such magnetism, because it is electrically produced, is called electromagnetisms.

【译文】由于这种磁性产生于电，所以称为电磁。

表示因果关系的英语复合句中，因果顺序灵活，在汉语中多数情况是原因在前，结果在后。

(3) 条件或假设关系的表述

例237 Not as much energy will be used as would be required if the circuit was closed for one hour instead of 10 minutes.

【译文】如果电路闭合时间不是10分钟，而是1小时，那么所消耗的能量就比所需要的要少得多。

另外，表示条件（假设）与结果关系的英语复合句中，条件（假设）与结果的顺序也不固定，在汉语中则是条件在前，结果在后。

6.6.2 长句翻译的基本方法

翻译长句时，首先要弄清楚原文的句法结构，找出整个句子的中心内容及其各层意思，辨清主次，然后分析各层意思之间的相互逻辑关系，把每个单句译成汉语，将这些汉语句子按照时间顺序和逻辑顺序，遵循汉语表达习惯进行调整和重新组合。最后，通读译文，进行润饰，避免出现欧化的句子。英语长句汉译时主要采用下列七种方法：直译法、包孕法、分译法、倒译法、拆离法、重组法和综合法。

6.6.2.1 直译法

例 238 Perhaps some prehistoric man found that piles of rocks across a stream would raise the water level sufficiently to overflow the land that was the source of his wild food plants and thus water them during the drought.

【译文】也许史前的人曾发现横贯河流的一块石头就能提高水位，足以淹没作为生长野生食用植物源泉的土地，而这样在干旱季节就能给植物浇水。

该句与汉语的行文习惯和逻辑基本一致，因此采用直译法翻译即可。

6.6.2.2 包孕法

这种方法是长句翻译中最基本的方法，就是汉译时将英文长句中的后置修饰成分都放在中心词的前面。这种译法可以使译文句子结构紧凑、整体感强。

例 239 The development of a practical transformer freed the utilities from the limitations imposed by the low voltage inherent in direct current.

【译文】实用变压器的研制使电力部门摆脱了直流电电压低所带来的局限性。

例中的 imposed by the low voltage inherent in direct current 是 limitations 的后置定语，但是汉语中的定语一般前置。因此，按照汉语的习惯，用包孕法来翻译，把过去分词短语放到 limitations 的前面作定语，译为“直流电电压低所带来的”。

例 240 On a natural water shed with many vegetal species, it is reasonable to assume that evapotranspiration rates do vary with soil moisture since shallow-rooted species will cease to transpire before deeper-rooted species.

【译文】对于一个有多种植物的自然流域，假设腾发速率确实随土壤水分的变化而变化是合理的，这是因为浅根植物将在深根植物前停止蒸腾。

例 240 中的过去分词短语 shed with many vegetal species 修饰先行词 water，翻译时采用包孕法，译作汉语的偏正关系名词短语“有多种植物的自然流域”。

例 241 Owing to the resistance, a current will heat the conductor along which it is passing.

【译文】由于有电阻，电流会使它所通过的导体发热。

例中的定语从句 which it is passing 修饰 conductor，由于这个从句不长，可以用包孕法译为：“它所通过的导体”。

6.6.2.3 分译法

汉语习惯用短句表达，而英语尤其是科技英语习惯多用长句，由于两种语言的句型结构上的差异，在科技英语翻译过程中，要把原文句子中复杂的逻辑关系表达清楚，往往要采用分译法。

分译法就是在句子太长、不合乎汉语表达习惯的时候，把句子割断，变成两个或三个句子来译。但是在什么地方割断很有讲究。一是在长句中有表示意思连接或转折的连词如“and、or、but”等处断开；二是在句中引导从句的关系代词或关系副词的地方断开；三是当长句中提到一个概念的多个方面时，可以通过重复主要词来断句。特别要注意的是断开的句子译成汉语后，要保持前后句意思上或文句上的联系，以保证原句意思的完整性。翻译时只需按原句顺序译出即可。由此可知，分译法就是按照意群将英文长句划分成几个部分，在汉译时，将英文长句化整为零，这样是为了符合汉语表达多用短句的特点。分译法可基本保留英语语序，形成流畅的表述，减少漏译。

例 242　The term quality is used here to include a number of important attributes dependent on design: appearance and visual coherence, which are significant factors in achieving and demonstrating quality; the compatibility of machines with their human and physical environment; reliability, especially as affected by maintenance facilities; safety, as influenced by control considerations, and the less specific quality of "wholeness", which is dependent on the designer's ability to optimize every detail in its relationship to the unit as a whole, and in which the synthesizing approach of industrial design makes a major contribution.

【译文】本书中所用的质量一词包括许多取决于设计的重要特性：外表与视觉连贯性、兼容性、可靠性、安全性和完整性。连贯性是获得显示质量的重要因素；兼容性是指机械与人和自然环境兼容程度；可靠性尤其受到维护条件的影响；安全性受到控制方面考虑的影响；完整性是不那么具体的特性，取决于设计师在细节与整体的关系中优化每个细节的能力，工业设计的综合手段对于完整性起着重要的作用。

此例很长，构成一个段落，但句子翻译的难度在于并列名词多且后面都跟有补充说明性的定语。如果在译文中将这些定语放在这些并列的名词前，译文就会十分杂乱，重点不突出。故重复并列名词，与它们的定语一起单独翻译成句子，仍按原来的顺序排列。这样，分开翻译不但没有破坏原文的意义，而且译文的条理十分清晰。

例 243　On the other hand, as we know, a good understanding has been obtained of wood and coal which consist chiefly of carbon, hydrogen and oxygen.

【译文】另一方面，就我们所知，对于木头和煤已经了解得很清楚，它们主要是由碳、氢和氧构成的。

这个句子的主句为 a good understanding has been obtained, of wood and coal 为 understanding 的定语，which 引导的宾语从句修饰 wood and coal。这个句子翻译时可在名词性从句 which consist of... 之前切断，化整为零。

例 244　Further interconnection of electric power systems over wide areas, continuing development of reliable automated control systems and apparatus, provision of additional reserve facilities, and further effort in developing personnel to engineer, design, construct, maintain, and operate these facilities will continue to improve the reliability of electric power supply.

【译文】在供电系统的大范围内进一步联网，不断发展完善可靠的自动化控制系统和设备，做好备用设施的供给工作，全力搞好设计、施工、维修和操作方面的技术培训，这都将不断提高供电的可靠性。

这一句共48个词，句子虽长，但结构并不复杂，其中前38个词都是主语部分。当然，主语中有许多并列成分和附加成分，其中，中心词分别为interconnection、development、provision、effort，句子的谓语是will continue，之后的不定式短语为宾语。层次结构明确，所以不难理解。

例245 The loads which a structure is subjected to are divided into dead loads, which include the weights of parts of the structure, and live loads, which are due to the weights of people, movable equipment, etc.

【译文】一个结构物受到的荷载可分为静载与活载两类。静载包括该结构物各部分的重量。活载则是由于人及可移动设备等的重量而引起的荷载。

此句采用分译法，化整为零，更符合汉语用小短句的习惯，且醒目易读。

例246 The resistance of any length of a conducting wire is easily measured by finding the potential difference in volts between its ends when a known current is following.

【译文】若电流已知，只要测出导线两端电位差的大小，便可容易地求出任一长度导线的电阻。

例句中when a known current is following是大前提，by finding the potential difference in volts between its ends是小前提，the resistance of any length of a conducting wire is easily measured是两个前提作用下的结论。弄清这层关系之后，就可以将原句化整为零，用汉语小分句的形式表述句义了。

例247 Fuel cells are devices that when a fuel such as hydrogen or hydrogen-rich compounds and oxygen is supplied to materials arranged like the anode and cathode of a conventional battery, combine to convert chemical energy directly into electrical energy.

【译文】燃料电池是这样一种装置：当供给氢（或富氢化合物）、氧反应物时，燃料电池中的阴阳两极就会像普通蓄电池中阳极和阴极一样发生化学反应，将化学能直接转变为电能。

句子的主干是fuel cells are devices，之后的that引导一个定语从句来修饰devices，从句中的谓语是combine to convert...；定语从句中有包含一个由when引导的时间状语从句，这个从句的主干结构为fuel...is supplied to...。由此，我们可以知道这个长句交代了：燃料电池是一种能直接将化学能转变为电能的装置；这两种能的转换的条件。第一个断句点是that前，第二个断句处是when引导的从句后。译文中根据需要分别用燃料电池重复这种装置所表达的意义。

例248 With the same number of protons, all nuclei of a given element may have different numbers of neutrons.

【译文】虽然某个元素的所有原子核都含有相同数目的质子，但他们含有的中子数可能不同。

上述的例子中，原文的一个简单句被译成了汉语的两个主谓句，由于with短语中的the same number of protons和句子谓语have different numbers of neutrons在意思上正好是平行且相对的，因此，我们可以把这个句子译成一个表示对比或转折关系的关联句。

例249 The unit of electrical energy is called the joule, which is equivalent to 10^7 ergs.

【译文】电能的单位是焦耳，1 焦耳等于 10^7 尔格。

which 引导一个非限制性定语从句，虽然在语法上是被修饰词 joule 的定语，但在语义上则是为了进一步解释说明，并没有明确的限定关系，因此，可以把它译成汉语中的一个后续性分句。

例 250 The discovery that electrical currents can be produced by magnetism is extremely important in the field of electricity.

【译文】磁能产生电流，这一个发现在电学上极为重要。

that 引导的是主语 discovery 的同位语从句，如果用包孕法，主语则太长，汉语势必会头重脚轻，这时可以先把同位语从句中的内容翻译出来，在后面用“这”或“那”等代词复指它。

6.6.2.4 倒译法

有些英语长句的表达顺序可能与汉语的表达正好相反，这种长句子在翻译时就得用倒译法。

例 251 Grounding every circuit, however, makes the system susceptible to excessive currents should a short circuit develop between a live conductor and ground.

【译文】然而，若将每一回路接地，当一个带电导体与地面之间发生短路时，会使系统对超过量电流太敏感。

should a short circuit develop between a live conductor and ground 是该句中的让步条件从句，为倒装语序。汉语的逻辑一般是先叙述条件，然后是结果，所以该句采用倒译法。

例 252 It is very interesting to note the differently chosen operating mechanism by the different manufacturers, in spite of fact that the operating mechanism has a major influence on the reliability of the circuit-breakers.

【译文】尽管操作装置对断路器的可靠性具有很大的影响，但注意不同的制造厂按不同形式选择操作装置是非常有趣的。

in spite of 是一个表示转折逻辑关系的介词，汉语对这种逻辑关系的表达习惯是“尽管/虽然……但是……”；而英文中这种逻辑关系的表述只需要一个连词或介词，而且顺序没有严格的规定。

例 253 The construction of such a satellite is now believed to be quite realizable, its realization being supported with all the achievements of contemporary science, which have brought into being not only materials capable of withstanding severe stresses involved and high temperatures developed, but new technological processes as well.

【译文】现代科学的一切成就不仅提供了能够承受高温高压的材料，而且也提供了新的工艺过程。依靠现代科学的这些成就，我们相信完全可以制造出这样的人造卫星。

例句由三部分构成：主句；作原因状语的分词独立结构；修饰独立结构的定语从句。根据汉语词序，状语特别是原因状语在先，定语前置，故从 which... 入手，再译出 its realization...，最后才译出 The construction... realizable。

例 254 Various machine parts can be washed very clean and will be as clean as new ones when they are treated by ultrasonic waves, no matter how dirty and irregularly shaped they may be.

【译文】各种机器零件无论多么脏，也不管形状多么不规则，若将其用超声波处理，都可以清洗得非常干净，甚至会洁净如新。

例句由三部分构成：主句（其中包含两个并列句）；可以转译为条件状语的时间状语从句；让步状语从句。根据三部分之间的语义关系和汉语表达习惯，采用倒译法：先表述放在最后的让步状语从句，再表述放在中间的时间状语从句，最后再表述主句。

6.6.2.5 拆离法

英语中有些长句很难用包孕法翻译，也很难用分译法顺译，这时只能将其拆开，这就要用到拆离法：将长句中的某些成分从句子主干中拆开，以利于译文的总体安排。

例 255 Somewhat surprisingly, the use of alternating current was at first opposed as dangerous by, for example, Edison who pointed out, accurately but irrelevantly, that it was used in the electric chair.

【译文】令人困惑的是，率先反对使用交流电的人竟是爱迪生。他举例说，电椅用交流电就很危险。电椅用交流电固然危险，但这与交流电的其他应用又有何干呢?

此例就是将原先长句中不容易处理的成分，如状语 somewhat surprisingly、for example 和 accurately but irrelevantly，主语补足语 as dangerous 从句子主干中拆开，单独翻译，从译文的整体布局上进行语序调整，从而使译文在忠实原文的前提下通顺流畅。

例 256 Many man-made substances are replacing certain natural products can not meet our ever-increasing requirements, or more often, because the physical property of the synthetic substance, which is the common name for man-made materials, have been chosen, and even emphasized, so that it would be of the greatest use in the fields in which it is to be applied.

【译文】人造材料通称为合成材料。许多人造材料正在代替某些天然材料，这或者是由于天然物产的数量不能满足人们日益增长的需要，或者往往是由于人们选择了合成材料的一些物理特征并加以突出而造成的。因此，合成材料在拟用的领域里将具有极大的用途。

在本句中，非限制性定语从句 which is the common name for man-made materials 夹在中间，如果要按原文的结构直译过来，译文就会前后不连贯。因此，翻译时将它抽出来分译，单独成句，放在句首，作为交代。由 so that 引起的结果状语从句也分译成独立的句子。其余部分基本上采用了直译。

例 257 The power plant had a capacity of four 250-hp boilers which supplied steam to six engine-dynamo sets and supplied power for lighting to 59 customers within an area of approximately 1 square mile.

【译文】此发电厂的容量为：4 台 250 马力的锅炉提供蒸汽给 6 台以蒸汽机为动力的发电机组。该电厂向大约 1 平方英里区域内的 59 个用户提供照明用电。

如果按照包孕法去翻译这句话，会被译成“这个发电站有……的容量”，句子拖沓冗长，还有可能出现漏译的情况。此外，也不能按照英文的语序顺译。可以把 which 这个非限制定语从句同主句先拆开，然后先翻译主句。如果谓语是 have，宾语后面有“of + 名词”，一般将该句的宾语转译为主语，将 of 后面的名词转译成宾语，也就是说按照拆离法，将句子的定语与所修饰的名词拆离，调整顺序，重新安排。

例 258 On account of the accuracy and ease with which resistance measurements may be made and the well-known manner in which resistance varies with temperature, it is common to use this variation to indicate changes in temperature.

【译文】我们都知道，电阻的大小是随温度而变化的，对电阻进行测量既精确又方便。因此，通常都用电阻的变化来表示温度的变化。

此例的总体结构是：表示原因的介词短语和主句。可以按照汉语习惯安排译文，即先原因后结果，采用顺译法。而介词短语中又含有两组并列的名词短语作宾语，并且这两组宾语后面又各有一个定语从句，这就是该句翻译的难点所在。分析一下原因状语中的各层在语义上的关系，可以考虑采用拆离法，先将名词短语 the well-known manner in which resistance varies with temperature 转译为句子“我们都知道，电阻的大小是随温度而变化的”，再将其前面的名词短语 the accuracy and ease with which resistance measurements may be made 也转译为句子“对电阻进行测量既精确又方便”。这样就把两个很长的名词短语译成句子，表达起来既顺口又准确。

6.6.2.6　重组法

重组法即打乱原文句序，然后仔细推敲，或按时间先后，或按逻辑顺序，有顺有逆、有主有次地改变句子结构，进行重新组合，综合处理。

例 259　There is another reason why AC is preferred to DC for long distance transmission.

【译文】为什么在长距离输电方面交流电优于直流电，这里还有一个原因。

这个句子如果用包孕法来处理，译文的基本结构是“还有另外一个……的理由”。这样一来，由于定语太长，句子变得拖沓，不太符合汉语的表达习惯。如果采用重组法，我们可以把 why 引导的从句译成一个与主句并列的分句，将主句放到从句后面翻译。

例 260　The tube consists of a short copper section followed by a longer steel section with a flange at the end.

【译文】管子由两段组成：前段短，是铜的；后段长，是钢的，末端带法兰。

汉语译文中将英语原文中所有介词词组和名词词组都转换成并列句，使译文符合中文的表达习惯。

例 261　Computer languages may vary from detailed low level close to that immediately understood by the particular computer, to the sophisticated high level which can be rendered automatically acceptable to a wide range of computers.

【译文】计算机语言有低级的也有高级的。前者比较繁琐，很接近于特定计算机直接能懂的语言，而后者比较复杂，但实用范围广，能自动为多种计算机所接受。

vary from... to 这种结构，在汉语中很难找到相对应的结构。若生搬硬套，译文就会显得生涩难懂。因此，在翻译过程中，经过正确理解全文的意思后，将原句结构顺序打乱，加以重组，采用增字、减字、调整译文顺序等技巧，将句中的形容词短语和定语从句拆开分译，并把作定语的过去分词和形容词短语结合在一句中。这样，整个译文就显得通顺地道了。

例 262　Vibration can lead to disintegration, for example, if heavily loaded studs are threaded into soft alloy, or friable plastics.

【译文】震动会导致解体。比如说，如果承重的螺栓拧进硬度不高的合金或易碎的塑料中时，就会发生这样的情况。

为了突出重点，对原文不长的句子也可以拆卸，通过添加和重复，译成两个独立句，如本例句所示。例句的语义重点是“震动会导致解体”，而从句则指出一个例证。如果译成：“震动会导致解体，比如说，如果承重的螺栓拧进硬度不高的合金或易碎的塑料中时”，或者“比如说，如果承重的螺栓拧进硬度不高的合金或易碎的塑料中时，震动会导致解体”，则前者不符合汉语中“条件在前，结果在后”的表达习惯，而后者好像整句话是在举例说明另外的现象，都不够妥当。所以，译文通过添加与文章重点相符的内容，即“就会发生这样的情况”并与从句一起组成另外一个独立句。这样，译文层次分明，重点突出。

例 263 The method normally employed for free electrons to be produced in electron tube is thermionic emission, in which advantage is taken of the fact, if a solid body is heated sufficiently, some of the electrons that it contains will escape from its surface into the surrounding space.

【译文】将固体加热到足够的温度时，它所含有的电子就有一部分离开表面，飞逸到周围的空间中，这种现象叫做热离子放射。电子管通常就是利用这种方法来产生自由电子的。

这是一个典型“先立主杆，后分枝叶”的复合句。主句比较简单，其主语有一个较长的过去分词短语作定语。但由 in which 引出的非限制性定语从句就比较复杂，因为它有一个 that 引导的同位语从句，而同位语从句中又含一个由 if 引出的条件状语从句和由 that 引出的定语从句。因此，在翻译时，必须要把句子顺序颠倒过来，先介绍现象，而后引出主题。这样，整个译文就符合汉语的表达习惯了。

例 264 The great difficulty of introducing radically new computer architectures which requires customers to rewrite most of their software excluded the possibility for those techniques to find their way to the commercial marketplace.

【译文】采用全新的计算机体系结构，势必要求用户改写其大部分软件。因此，难以付诸实现，这就排除了这种技术进入商品市场的可能性。

例中 of 引导的介词短语做后置定语修饰主语 difficulty，但由于 of 引导的介词短语结构是含有一个限制性定语从句的复合句，难以直译，因此，在翻译时，必须分拆原句，重新组合：把定语从句拆出分译，再把 great difficulty 抽出来译为“难以付诸实现”，独立构成一句。这样，不仅圆满地解决了上述的诸多困难，而且也使译文更加通顺流畅。

6.6.2.7 综合法

综合法指同时采用上述的两种或三种方法。

例 265 This hope of “early discovery” of lung cancer followed by surgical cure, which currently seems to be the most effective form of therapy, is often thwarted by diverse biologic behaviors in the rate and direction of growth of the cancer.

【译文】人们希望“早期发现”肺癌，随之进行外科治疗，因为外科治疗目前可能是效果最好的办法。然而，由于肺癌生长速度和生长方向等生物学特征有很大差异，早期发现的希望往往落空。

例句翻译时译成两句，第一句顺译，第二句倒译。

例 266 There is enough of a difference here to indicate that one must look at the foreman’s job in terms of what his situation is, whom he has to motivate and what opportunities he has to do——before deciding what sort of supervisor training is best for him.

【译文】这里的差别足以证明：在决定何种管理训练对工长最有用之前，人们必须从工长所在的位置，他需要促动什么人和他有哪些进行促动的机会等方面先对他的工作进行一番考察。

例句翻译时冒号之前是顺译，冒号之后是倒译。

在翻译实践中，以上所提到的长句翻译方法并不是孤立的、绝对的。任何方法的运用都是为了使原文简明通顺，易读易懂。作为译者，必须学会将各种方法有机地结合起来，要以完整、准确、通顺地表达出原文的意义为翻译的最终目的。

6.7 非人称主语句的翻译

科技英语的一个显著特点就是非人称主语句多。科技英语的主要目的在于阐述科学事实、科学发现、试验结果等，是为了说明科技活动所带来的结果、证明的理论或发现的科学现象和科学规律，因此，往往没有人称。非人称主语句，就是其主语是无生命的事物或抽象名词。由于汉语中很少采用这种句式，因此，汉译时为了使译文简洁规范，往往需要根据句子的逻辑关系，做一些变通处理。

6.7.1 译成人称主语句

例267 The successful launching of China's first experimental communications satellite, which was propelled by a three-stage rocket and has been in operation ever since, indicates that our country has entered a new stage in the development of carrier rockets and electronics.

【译文】中国已成功地发射了第一颗实验通讯卫星。这颗卫星是由三级火箭推动的，一直运转正常。它标志着我国在发展运载工具和电子技术方面进入了一个新阶段。

例268 The efforts that have been made to explain optical phenomena by means of the hypothesis of a medium having the same physical character as an elastic solid body led, in the first instance, to the understanding of a concrete example of a medium which can transmit transverse vibrations, and at a later stage to the definite conclusion that there is no luminiferous medium having the physical character assumed in the hypothesis.

【译文】为了解释光学现象，人们曾试图假定有一种具有与弹性固体相同的物理性质的介质。这种尝试的结果，最初曾使人们了解到一种能传输横向振动的具有上述假定所认为的那种物理性质的发光介质。

以上两例都在翻译中适当增加了人称代词作主语，这个主语一般是泛指人称代词或逻辑主语，否则句子中就少了施动者。

6.7.2 将主语译成短语

例269 A command of knowledge of computer science is of great help to the study and research in all fields.

【译文】掌握一些计算机知识对各领域的研究是很有帮助的。

例中将主语译为动宾短语，仍作主语。

6.7.3 译成汉语的后续分句

例 270 Hydraulic system malfunctions occasionally call for tests such as foaming tendency, water separation, and rust or oxidation resistance.

【译文】有时液压系统工作反常，我们就需要做各种试验，如：冒泡试验、水分离试验和防蚀或防氧化试验。

例 271 From 1970 onwards, advances in large scale Integration (LSI) techniques, and availability of inexpensive microprocessors led to the possibility of producing an all digital, selfcontained real time analyzer.

【译文】自 1970 年起，大规模集成电路技术不断发展，廉价的微处理器相继出现，于是人们能够生产出全数字化自容式实时分析器的可能性。

例 272 The discovery of new species will lead to new drugs.

【译文】发现了新的物种，我们就可以研制新药。

例 273 The application of computers brings a tremendous rise in labor productivity.

【译文】使用了电脑后，劳动生产率突飞猛进。

以上例句在翻译时，主语都视具体情况转译成了汉语的一个小分句，通常放在汉语译文的前半部分，例 270 ~272 在后续分句中增译泛指人称代词作主语，而例 273 则没有添加人称代词。

6.7.4 将作宾语的人称代词译成主语

例 274 The development of a practical transformer enables the utilities to break through the limitations imposed by the low voltage inherent in direct current.

【译文】有了实用变压器，电力部门就可以摆脱直流电电压低所带来的局限性。

例 275 The data edit controls located at the left-hand side of the cabinet consisting of 16 separate channel adjustments together with the data edit switch and data gain control, enable the operator to modify the output values.

【译文】操作者可借助位于壳体左侧的一些数据编辑控制器修正输出值，这些控制器包括 16 只独立的通道调节器，以及数据编辑开关和数据增益控制器。

当无人称主语句的谓语为 enable sb. to do sth. 时，可以把宾语译成主语。例 274 和例 275 的宾语 the utilities 和 the operator 对应译文的主语，enable sb. to do sth. 结构中的 sb. to do sth. 翻译时都转译成一个完整的主谓句“电力部门就可以摆脱直流电电压低所带来的局限性”和“操作者可借助位于壳体左侧的一些数据编辑控制器修正输出值”。

翻 译 练 习

1. A step-down transformer can reduce voltage to whatever is desired.
2. Attention is necessary to prevent the electronic instrument from damage.
3. This paper aims at discussing the application of nuclear energy.
4. The increase in pressure with depth makes it difficult for man to go very deep far below the water surface.

5. Some operating conditions could produce elevated casing temperature presenting a fire hazard.

6. With the brain at work, all the parts of the body work most smoothly.

7. Atoms are too small to be seen even with a powerful microscope.

8. We find almost all physical properties of matter influenced by heat.

9. The study found a direct link between the amount of fish in the diet and the rate of death from heart disease.

10. This phenomenon accounts for a lower density of ice than that of water.

11. We could use two resistors in series, or we could increase the value of the present single resistor.

12. An alternating current has not a constant direction, and it has no constant magnitude.

13. The turns of line must be insulated from each other and from the iron, or the current will flow through short-circuits instead of flowing around the coil.

14. Positive frequencies are produced by anticlockwise rotation and negative frequencies by clockwise rotation.

15. Look at crystals of table-salt under a magnifying glass and you will see that each crystal is a cube with 6 faces.

16. There exist neither perfect insulators nor perfect conductors, for all substances offer opposition to the flow of electric current.

17. The material first used was copper for it is easily obtained in its pure state.

18. The average kinetic energy of the molecules is increased and the temperature is raised.

19. The number of protons is equal to that of electrons and the whole atom is electrically neutral.

20. Displacement is a vector and time is a scalar.

21. The longer the conductor is, the greater the resistance is.

22. All losses of energy in heat engines, great as they are, do not in any way contradict the law of conservation of energy.

23. The sort of lubricant which we use depends largely on the running speed of the bearing.

24. Nuclear weapons, which are after all created by man, certainly will be eliminated by man.

25. Electricity which is passed through the thin tungsten wire inside the bulb makes the wire very hot.

26. Such atoms are so constructed that they lose electrons easily.

27. Though the cost of the venture would be immense, both in labour and power, many believe that iceberg towing would prove less costly in the long run than the alternative of desalination of sea water.

28. It also plays an important role in making the earth more habitable, as warm ocean currents bring milder temperatures to places that would otherwise be quite cold.

29. The subject was discussed because of a general belief that very little feedback had been provided from the 1976 High Reynolds Number Research Workshop.

30. The hydrogen produced by electrolysis is nearly pure, though rather more expensive than that obtained by the thermal cracking process.

31. Because energy can be changed from one form into another, electricity can be changed into heat energy, mechanical energy, light energy, etc.

32. Electric charges work when they move.

33. As the sphere becomes larger, the waves become weaker.

34. The side of Mercury which is turned away from the sun remains in eternal darkness, with a temperature only a few degrees above absolute zero.

35. An "alloy" steel is one which, in addition to the contents of carbon, sulphur and phosphorus, contains more than 1 percent of manganese, or more than 0.3 percent of silicon, or some other elements in amounts not

encountered in carbon steels.

36. The control system may include a man stationed in the power plant who watches a set of meters on the generator output terminals and makes necessary adjustments manually.

37. The arteries and veins do not allow anything to pass through their walls, but the capillaries, which are the smallest blood vessels, allow water and other small molecules to pass into the tissues.

38. When winds blow particles against a large rock for a long time, the softer layers of the rock are slowly worn away.

39. From an economic point of view, integrated circuits mean much lower costs through the use of automated mass production methods.

40. An infrared system could be useable in both anti-air and anti-ship engagements but its inherent disadvantages related to some dependence on optical visibility and to sensibility to interference from natural or man-made sources make it less attractive in the surface than in the air role.

41. From what is stated above, it is learned that the sun's heat can pass through the empty space between the sun and the atmosphere that surrounds the earth, and that most of the heat is dispersed through the atmosphere and lost, which is really what happens in the practical case, but to what extent it is lost has not been found out.

42. Captured documents which we have obtained from individuals who had been infiltrated through this corridor plus prisoner-of-war reports that we have obtained in recent months led us to believe that the volume of infiltration has expanded substantially.

43. Despite the backwardness (in the communications) ——or perhaps because of it——China hopes to leapfrog into the digital era by bypassing many of the costly transitional technologies that industrial nations are now seeking to replace with more advanced digital systems.

44. He had attached a wheel with a notched rim to the clock, connected this with the telegraph line, and so arranged it that every time the hour was reached, the right number of dots was automatically sent along the line by the turning of the wheel.

45. If parents were prepared for this adolescent reaction, and realized that it was a sign that the child was growing up and developing valuable powers of observation and independent judgment, they would not be so hurt, and therefore would not drive the child into opposition by resenting and resisting it.

46. It was our view that the United States could be effective in both the tasks outlined by the President-that is, of ending hostilities as well as of making a contribution to a permanent peace in the Middle East——if we conducted ourselves so that we could remain in permanent contact with all of these elements in the equation.

47. The second aspect is the application by all members of society from the government official to the ordinary citizen, of the special methods of thought and action that scientists use in their work.

48. In a potentially important reversal of roles, a mainland Chinese electronics company based in Shenzhen has kickstarted a multi-billion dollar investment in Hongkong's laggard high-technology sector by signing a contract with the Hongkong Government to take a site for a planned US $312 million state-of-the-art semiconductor plant.

49. A complete absence of oxygen and water in the moon makes it a dead world with no signs of life.

50. To count the stars in the Milky Way by ordinary methods would take more than a lifetime.

51. The kill probability against a tank has been calculated between 30% and 40%.

52. There is the challenge, more clearly defined than ever before, to scientists to apply the results of science and technology for the benefit of mankind.

53. There is still the faith that ordinary men are greater than the powers of nature or the mechanisms of man's hands and will master them all in the end.

54. Adherence to one One-China Principle is the basis and prerequisite for peaceful reunification china.

第7章　语态、时态和语气的翻译

7.1　被动语态的翻译

语态是用动词表示主语与谓语之间的关系的一种形式。英语语态分为主动语态和被动语态两种。英语的主动语态与汉语的表达方式接近，符合中国人的表达习惯。被动语态在科技英语中使用得非常广泛。据统计，英语科技文献中50%以上的谓语动词是用被动语态表述的。科技英语之所以如此频繁地使用被动语态原因有三：第一，被动结构比主动结构主观色彩更少，科技论著重客观事实，正需要这种特性。第二，被动结构更能突出主要论证及说明对象，更加引人注目。第三，在很多情况下，被动结构比主动结构更简洁。

与英语相比，汉语虽然也有被动语态，但远不如英语那样广泛。同样一个概念，英语用被动语态表达，而汉语却用主动语态表达。因此，在翻译英语被动语态的句子时，不必恪守原句的语态，而应该设法摆脱这种被动的局限性，灵活采用多种翻译技巧，运用各种修辞手段。只有这样才能使译文既在科技内容上忠实于原文，又在语言形式上臻于通顺，符合汉语的表达方式。现就英语被动语态句的翻译方法介绍如下：

7.1.1　译成汉语的被动句

7.1.1.1　直接加译“被”字

对于某些特别强调被动动作或特别突出被动者行为的英语被动句，在译成汉语时可以在谓语动词前直接加译“被”字。

汉语里表示被动的方式很多，除了用“被”字以外，还可以使用与被动意义相同相近的字词，如“叫”、“让”、“受”、“靠”、“为”、“挨”、“给”、“用”、“给予”、“加以”、“为……所……”等。

例1　If the NDA is destroyed, the cell cannot divide, and will die.

【译文】如果NDA被破坏，细胞不能分裂，就会死去。

例2　The virus is quickly killed upon exposure to the air.

【译文】这种病毒一接触空气就会被迅速杀死。

例3　Once the impurities have been removed, the actual reduction to the metal is an easy step.

【译文】杂质一旦被除掉，金属的真正还原就容易了。

7.1.1.2　加译含有被动意义的其他汉字

在不改变原文语序的情况下，有些英语被动句可加译“受”、“用”、“由”、“让”、“给”等，表示被动的意义。

例 4 The γ-rays are not affected by an electric field.

【译文】γ 射线不受电场影响。

例 5 If the work piece is grabbed directly, it warps due to the body temperature.

【译文】如果用手直接抓工件，工件就会因受到体温而发生变形。

例 6 If a body is acted on by a number of forces and still remains stationary, the body is said to be in equilibrium.

【译文】如果一个物体受到若干力的作用仍然保持静止，就可以说该物体处于平衡状态。

例 7 An oxidation number may be assigned to each atom in a substance by the application of simple rules.

【译文】应用一些简单的规则，可以给一种物质里的各原子指定氧化值。

例 8 The second group is composed of compounds derived from or related to benzene C_6H_6.

【译文】第二族是由苯 C_6H_6 衍生的或由与苯有关的化合物组成。

例 9 The crops were washed away by the flood.

【译文】庄稼让洪水冲跑了。

此类句子中，在动词前面，“由”、“让”、“给” 等字后面有一个名词作为该动词的逻辑主语。这是与直接译成被动语态句子的不同之处。

7.1.1.3 加译复合动词

在译文中添加 “靠”、“为”、“挨”、“遭受”、“给予”、“加以”、“为……所……” 等动词。

例 10 The airplane is supported by the wings; it is propelled by the power plant; it is guided by its control surfaces.

【译文】飞机靠机翼支承，靠动力装置推进，靠操纵面导向。

例 11 This extraction rate was confirmed in batch tank tests.

【译文】这一提取速度已为分批槽内试验所证实。

例 12 Other advantages of our invention will be discussed in the following.

【译文】本发明的其他特点将在下文中予以讨论。

例 13 These problems must be solved before the test starts.

【译文】这些问题必须在试验开始前加以解决。

例 14 North China was hit by an unexpected heavy rain, which caused severe flooding.

【译文】华北地区遭受了一场意外的大雨袭击，引起了严重的水灾。

例 15 The hypothesis was not accepted by most chemists until the 1970s.

【译文】这一假设直到七十年代才为大多数化学家所接受。

此类句子在汉译时不用改变原文语序，可以在谓语动词前加译 “遭受”、“给予”、“为……所……” 等词，把英语被动句译成汉语被动句。

7.1.1.4 译成汉语的判断句

可将被动语态译为判断句 “……的是……” 或 “……是……的”。

例 16 The first car driven by one of these engines was seen on the roads in 1894.

【译文】第一辆用内燃机提供动力的在马路上行驶的汽车是 1894 年出现的。

例 17 Isolated columns or stanchions are normally supported on square concrete foundation bases.

【译文】独立的圆柱或立柱通常是支撑在正方形的钢筋混凝土基础上的。

例 18 The shell parts of reactor pressure vessels have been often fabricated with formed plates welded together.

【译文】反应堆压力容器的壳体常常是用成型钢板焊接而成的。

例 19 Associated with the turn knob are two micro switches, which are closed when the knob is in the detent position.

【译文】与转向旋钮连接的是两个微型开关，它们在转向旋钮处于止动位置时是闭合的。

例 20 In an ordinary atom the number of protons are the same as that of electrons. Clustered together with these protons are neutrons.

【译文】在普通原子里，质子数和电子数相等。与这些质子集结在一起的是中子。

"……的是……"或"……是……的"句型是汉语的判断句型，起强调作用，适用于说明或表示某一事实或状态。

7.1.2 译成汉语的主动句

某些情况下，英语的被动句可以译成汉语的主动句：第一，英语中的某些被动句用来表示某种状态或某种动作的结果，而动作的实施者不太明确或很难指出，这类句子实际上并不强调被动的意义。第二，有些被动句其动作的实施者虽很明确，但意在突出动作的承受者或强调某种行为或动作的结果，因而把这些承受者或结果置于主语的位置上以示醒目或以便表达，这类句子实际上也不强调被动的意义。第三，有些被动句具有典型的被动意义，并且带有由 by 引导的行为主体。第四，有些被动句的表达法是出于行文上的需要，或是一种习惯表达法。

英语被动句译成汉语主动句有下列几种情况：

7.1.2.1 原文的主语仍译作汉语的主语

例 21 The experiment will be finished in a week.

【译文】这项试验将在一周后完成。

例 22 The robot sifter has been put into use.

【译文】自动筛机已投入使用。

例 23 Account should be taken of the low melting point of this substance.

【译文】应该考虑这种物质的低熔点。

例 24 Accompanying the visible light, a great deal of invisible radiation, or radiant heat, is emitted.

【译文】大量的不可见的辐射，伴随着辐射热释放出来。

例 25 Several approaches to the problem of slag or deoxidation-scum removal are being tried.

【译文】熔渣或脱氧浮渣清除问题的几项解决方案正在试验中。

例 26 The AIDS virusis only transmitted by blood and other body fluids.

【译文】艾滋病毒只能通过血液和其他体液传播。

以上例句在汉译时均按照原句的意思和译文的表达习惯，将句子译成汉语的主动句，原句的主语在译文中仍作主语。

7.1.2.2 原文其他成分改译成主语

有些英语被动句的主语无法像以上例句那样，译成汉语时仍作主语。我们可把原句中的某一其他成分改译作汉语的主语，而把原主语改译为其他成分。译作主语的大多是由 by 引导的行为主体，或句中的某一状语成分。

例 27 Heat and light can be given off by this chemical change.

【译文】这种化学反应能放出热和光。

例 28 More and more iron and steel will be produced in our country.

【译文】我国将生产越来越多的钢铁。

例 29 Defects in welds can be detected by nondestructive testing methods.

【译文】无损检验法可探查焊缝内的缺陷。

例 30 Communications satellites are used for international living transmission throughout the world.

【译文】全世界都将通讯卫星用于国际间的实况转播。

例 31 Three-phase current should be used for large motors.

【译文】大型电动机应当使用三相电流。

以上例句译法的特点是，把原文中作状语的介词结构中的宾语译成主语，同时把原句的主语译成宾语。

7.1.2.3 译成汉语的无主句

英语中许多被动语态的句子、往往可译成汉语的无主句，这时英语被动句中的主语就译成了汉语无主句中的宾语。这种译法在科技文章中颇为常见，因为一般不需要指出动作的发出者。有时还可在句中原主语之前加“把”、“将”、“使”、“给”、“对”等词。

例 32 The resistance can be determined provided that the voltage and current are known.

【译文】只要知道电压和电流，就能确定电阻。

例 33 Air resistance must be given careful consideration when the aircraft is to be manufactured.

【译文】要制造飞行器，必须仔细考虑空气阻力问题。

例 34 Reinforced concrete roofs are often constructed nowadays, especially where there is a risk of fire.

【译文】从防火要求考虑，目前多采用钢筋混凝土的屋顶。

例 35 Where tool breakage is a problem, negative rake and stronger grade carbides are recommended.

【译文】在工具容易损坏的地方，最好采用负刃倾角和高强度的碳合金。

例 36 Whenever work is being done, energy is being converted from one form into another.

【译文】凡做功，都是把能从一种形式转换成另一种形式。

例 37 In order that an inflammable gas may burn in air, it is raised to the ignition temperature.

【译文】为了使可燃气体在空气中燃烧，就要把它提高到燃点温度。

例 38 When such motors are used, they are usually built into the equipment by the machinery manufacturer, as in portable tools, office machinery, and other equipment.

【译文】这类电动机的使用通常是由机械制造商把它们直接安装在设备中，如便携式工具、办公机械及其他设备。

例 39 Sand may be carried many miles by the wind.

【译文】风可以把沙土带到许多英里以外的地方。

翻译此类句子时应注意：在没有 by 引导的被动句中，可将“把”字放在原文句子的主语之前，即将原英语被动句的主语转译成“把”字的宾语，整个句子译成汉语中的无主句，如例 36、例 37。当原文为含有 by 引导的被动句时，通常是将 by 后面的名词译成句子的主语，整个句子译成带主语的“把”字句，如例 38、例 39。

7.1.3 习惯译法

7.1.3.1 it 作形式主语的被动语态

在科技英语中，“it + 被动态谓语 + 主语从句”的句型非常多。这种句子常常译为无主句或采用加译主语的译法。

例 40 It is generally accepted that irradiation is destructive to human tissues.

【译文】一般认为辐射对人体组织是有害的。

例 41 It is estimated that the human eye can distinguish 10 million different shades of colors.

【译文】据估计，人的眼睛能分辨出一千万种不同的颜色。

例 42 It was thought at one time that compound of carbon were only produced in living organisms.

【译文】人们曾一度认为，碳化合物只有在活的有机体中才能制造。

例 43 It is found that not all metals conduct electricity equally well.

【译文】人们发现，并非所有金属都能同样好地导电。

例 44 It is said that in industry certain mixtures of metals or alloys have been considerably developed.

【译文】据说，在工业中某些金属混合物即合金已经大大地发展了。

类似的句型在习惯上有固定的译法。如：

It must be admitted that... 必须承认……

It will be seen from this that... 由此可见……

It is stated that... 据说……

It is demonstrated that... 据证实……

It is assumed that... 假定……

It is mentioned that... 上文提及到……

It cannot be denied that... 无可否认……

7.1.3.2 由 as 引出主语补足语的被动语态

在科技英语中，有不少由 as 引出主语补足语的短语，它们在翻译成汉语时亦有固定的译法。

例 45 Heat is regarded as a form of energy.

【译文】热被看作是能的一种形式。

例 46 When a space man is travelling in space or orbiting round a planet, his feeling of weight which the force of gravity produces is gone. This condition is considered as weightlessness.

【译文】宇航员在太空飞行或环绕行星运行时，对由引力产生的重量感觉不出来，这种状态被认为是失重。

类似的句型在习惯上有固定的译法。如：

be referred to as　指为……，叫做……

be spoken of as　称为……，说成……

be regarded as　看成……，当作……

be treated as　当作……

be thought of as　认为……，当作……

be defined as　定义为……，定义是……

7.2 时态的翻译

科技英语中常用的时态有六类：一般现在时、一般过去时、一般将来时、一般过去将来时、进行时和完成时。英语的时态变化是依靠使用不同的助动词和通过动词词形的变化来实现的。而在汉语里，因为没有助动词及词形变化，只能利用时间副词“现在”、“将来”、“正在”、“曾经”、“经常”等及助词“着”、“过”、“了”等来表示不同的时间概念。另外，科技英语中有些表示时态的谓语动词并不表示时间概念，而是出于文章整体的需要，所以有时候没有必要一一译出。

7.2.1 一般现在时的译法

一般现在时是科技英语文章中最常用的时态。一般现在时的译法比较简单，除了在表示主语的特征时需要在动词前面加一个“能”、“可”或“会”字外，可直接译出。

例 47 Control of the refrigeration process is simplified by the use of a liquid sub-cooling control valve.

【译文】冷冻过程的控制可采用一个液体过冷控制阀而得到简化。

例 48 Electricity plays a very important role in our daily life.

【译文】电在我们日常生活中起着十分重要的作用。

7.2.2 一般过去时的译法

一般过去时的译法比较简单，一般不需要添加什么副词或助词来表示过去时。这是因为汉语习惯上不需要明确表示动词的时态，根据上下文，或借助句子里的时间状语，便可表达过去的时间。但有时为了更准确地翻译，或为了强调起见，也可在动词前后添加“已”、“曾”、“过”、“了”等字，或在句首添加“以前”、“当时”、“过去”等时间副词。

例49 Preliminary hormone studies did not reveal any striking changes in this tissue.

【译文】初步的激素研究未曾显示这一组织有任何显著改变。

例50 Twenty years ago the rolling mill was the centre of attention since the main interest was in meeting the steadily rising demand.

【译文】二十年以前轧机曾经是注意的中心，因为当时主要的兴趣在于满足稳定增长的需要。

例51 He developed a computer model based on the data accumulated in these studies.

【译文】他根据研究中积累的数据研制了一个计算机模型。

7.2.3 一般将来时的译法

一般将来时一般表示将要发生的动作与状态或将来经常发生的动作。翻译这种时态时，大都可以在动词前面添加"将"、"要"、"会"等字。

例52 If the resistance of the circuit is high, the current will decrease rapidly, but a high induced e. m. f. will result.

【译文】如果电路的电阻高，电流就会迅速减小，但会产生强的感应电动势。

例53 Fossil fuels will continue to act as the principal fuel source in the first quarter of the 21st century.

【译文】矿物燃料在21世纪的前25年里将仍然是主要的燃料来源。

以上两句译文中的will兼有情态动词的意思。另外，be going to（就要，即将）和be to（将要，需要）这两个短语（后面都要跟动词原形，构成复合谓语）也都表示将来的动作，可以译成"将"。

例54 We are going to launch another artificial satellite of earth next week.

【译文】下周我们将发射另一颗人造地球卫星。

例55 She is going to work at a rolling mill.

【译文】她即将到轧钢厂工作。

7.2.4 过去将来时的译法

这种时态常用在主句动词为过去时态的名词性从句（主要是宾语从句）中，翻译时常在动词前面添加"将"、"要"等字，和一般将来时译法相同。

例56 We were sure that special radio instruments would provide automatic take-off and landing of airplane in the future.

【译文】我们那时确信专门的无线电仪表以后会为飞机的自动起飞和着陆提供条件。

例57 They said that they would check up and repair these machines.

【译文】他们说过要检查和修理这些机器。

例58 I knew that I should take part in an important meeting next week.

【译文】当时我知道我将在后一周参加一个重要的会议。

7.2.5 进行时态的译法

进行时表示某一时间正在进行的动作。译成汉语时，通常要添加"正"、"在"、"正在”等字样，有时也可在动词后面加上助词“着”字。

例 59 Contemporary natural sciences are now working for new important breakthroughs.

【译文】当代自然科学正在孕育着一个新的重大突破。

例 60 Even before the Second Five-year Plan, China was already producing all kinds of lathes, machines, apparatus and instruments.

【译文】甚至在第二个五年计划之前，中国就已经在生产着各种车床、机器、器械和仪表。

例 61 Since 1964, Alcan has been using digital computers to help control and optimize the output of its various processes.

【译文】自 1964 年以来，加拿大铝业公司一直在利用数字计算机来帮助控制公司的各种生产过程和使该过程的生产最佳化。

例 62 Important research had been going on both sides of the Atlantic, mainly in connection with the mining industry.

【译文】在大西洋两岸过去一直在进行着重要的研究，后者主要与采矿工业有关。

从例 61、例 62 的译文中可以看出，在翻译完成进行式时，常须在动词前添加“一直在”字样，借以强调事情的延续性。

7.2.6 完成时态的译法

在科技英语中比较常见的是现在完成时它表示这样几种情况：第一，到现在为止已经或刚刚完成某种与现在有关的动作。第二，过去已经开始而持续到现在的动作（常与 since、for、in the past 等时间状语连用）。第三，在时间或条件状语从句中也常用将来完成时。翻译完成时态时可在动词前面添加时间副词“已（经）”、“曾（经）”，在动词后面添加助词“了”、“过”或“过……了”。

例 63 In modern industry drop forging has been finding wide use.

【译文】在现代工业中落锤锻造一直得到广泛应用。

例 64 In recent years scientists have found chemicals that affect mood, memory, and other happenings of the mind.

【译文】近年来，科学家们已经发现了一些会影响情绪、记忆以及其他精神活动的化学物质。

例 65 The principles of aluminum foil and strip manufacture have been the subject of a series of investigations recently.

【译文】铝箔和铝带的制造原理近来已成为一系列研究的主题。

7.3 语气的翻译

英语的语气在功能上与汉语差异不大，都是表示说话人对某一事件的看法或态度，但表现方式不同。英语是靠改变句子的结构和语序，使用某些特定的词语，包括助动词、动词、名词、形容词、副词和改变动词时态等方式来表现的；而汉语则主要是利用语气词“啊、啦、了、吧、呢”等表示。英语里有四种语气：陈述语气、疑问语气、祈使语气和虚拟语气。下面着重讨论一下祈使语气和虚拟语气中动词的译法。

7.3.1 祈使句的译法

祈使语气的特点就是无主语，可用来表示请求、要求、命令、劝告等，在科技英语中更常用来表示建议和指示，特别是在工艺操作规程和设备说明书中大量使用祈使句。此外，还可用来表示条件或设想，一般译成无主句。英语中的祈使语气在译成汉语时，可根据句子的不同情况增加“要”、“请”、“应”、“须”、“试”、“应该”、“千万”、“一定”、“务必”等词，有时也可以译成“把”字句。

例66 Don't forget to open the switch.

【译文】不要忘记拉开电闸。

例67 Let R denote the resultant of all forces.

【译文】设R表示所有的合力。

例68 Place a clean iron part in the solution of copper sulfate, and the part will be coated with red copper.

【译文】把一个干净的铁制件放在硫酸铜溶液中，它就会涂上一层紫铜。

例69 Note that increasing the length of the wire increases its resistance.

【译文】应注意：增加导线长度就会增加导线的电阻。

例70 Calculate the unknowns in the following.

【译文】试计算下面的未知数。

7.3.2 虚拟语气的译法

英语的虚拟语气是通过谓语动词的词形变化来实现的，而在汉语里，因为没有词形变化，只能通过一些连词和副词来表达“虚拟”的含义。因此，有些真实条件句和虚拟语气在翻译上很难体现出来。

例71 If it is used in this way, it will break.

【译文】如果这样使用的话，它就会断裂。

例72 If it were used in this way, it would break.

【译文】如果这样使用的话，它就会断裂。

例73 What happens if we place a coil of wire in the circuit?

【译文】我们在线路中放一个线圈，会发生什么情况呢?

例74 What could happen if we placed a coil of wire in the circuit?

【译文】我们在线路中放一个线圈，会发生什么情况呢?

例71和例73是真实条件句，例72和例74是虚拟语气，在英文中可以根据时态的使用来区分，但在汉译时却不能通过谓语动词时态的变化来区分。

7.3.2.1 虚拟条件句的译法

汉语常用“假使、要是、倘若、如果”等连词和“本来、就、会、也许”等副词表示虚拟条件。为了表达过去的时间，还常用“早、已经、曾经、以前”等副词。

例75 If metals had no such properties as ductility and malleability, computers could not be made.

【译文】如果金属没有延性、展性这些特性的话，就无法制造计算机。

例 76 These several compounds could not have been formed if the chemical reaction had been stopped.

【译文】倘若这一化学反应曾经停止过，就不能形成这几种化合物了。

例 77 If facts had been collected one week earlier, we should have had more time to study them.

【译文】如果一星期前就收集到了这些情况，我们就有更多时间进行研究。

例 78 If you had used the right method, you could have solved the problem of current leakage.

【译文】如果你当时使用了正确的方法，你本可以解决漏电问题。

例 79 We could have completed this experiment on fatigue life.

【译文】我们本来是能够完成这项有关疲劳寿命的实验的。

7.3.2.2 虚拟比较句的译法

汉语常用“好像”、“好像……样子”、“宛如”、“似乎”等词表示虚拟比较。

例 80 The machine is very large as if it were a building.

【译文】这部机器很大，似乎像一幢大建筑物。

例 81 The whole weight of a body acts as if it were concentrated at a single point, this point being called C. G.

【译文】物体的整个重量似乎是集中在一个点上而在起作用，而这一个点叫做重心。

例 82 Things far off look as if they were smaller than they really are.

【译文】远处的物体看起来好像比它们的实际尺寸小些。

7.3.2.3 虚拟让步句的译法

汉语常用“如果”、“除非”、“倘若”、“万一”、“假使”等连词表示虚拟让步。

例 83 If something should go wrong, the signal lamp would light up.

【译文】万一发生什么事故，信号灯就会照亮。

例 84 Without electric pressure in a semiconductor, the electron flow would not take place in it.

【译文】要是半导体没有电压，其内部就不会产生电子流。

例 85 But for electronic computers, there would have been no artificial satellites or rockets.

【译文】如果没有电子计算机，就不会有人造卫星和火箭了。

例 86 Unless this part had been overheated, it would not have failed.

【译文】除非这个零件过热了，否则它就不会损坏。

例 87 Suppose that you could magnify one atom so that it become as large as a big room.

【译文】假使你可以把一个原子放大得像一个大房间一样大。

翻译练习

1. Thus electricity is seen to be a form of energy by which power may be taken up at one place and rapidly——and generally very efficiently——delivered at another.

2. The process is similar to that performed at the transmit end, and the carriers are suppressed within the balanced modulation.

3. Pure oxygen must be given to patients in certain circumstances.

4. Because it is not heavy for the power it gives, a petrol engine is still used in small aircraft.

5. Innovative glass technology has been driven by increasing concern with climate changes, energy conservation, and urban sustainability.

6. A lot has been found out about the journeys of migrating birds by marking the birds with aluminum rings put on one leg.

7. Pipes are often corroded by the oxygen in the air.

8. Friction is also much diminished by proper lubrication.

9. There are a number of significant top-level issues that must be addressed if a cooperative approach to human space exploration is to be pursued.

10. This latter decision was implied in 2004 by the budget projections released together with the VSE plan, and has never been explicitly countermanded.

11. Meteorologists are currently trying to determine what proportion of the continent's precipitation is produced in this way.

12. At one time, it was thought that nothing could possibly live very deep down in the oceans.

13. Assuming radiation losses to be 5%, calculate the amount of steam required.

14. They believe that they will bring to discussions of their participation in the VSE more bargaining power than has been in the case historically.

15. In fact, the hole appears to move in response to an applied electric field, as though it were a particle exhibiting both a positive charge and a positive mass.

16. The ISS mechanism has worked, since the station is now reaching its final stage of assembly.

17. Lunar surface systems make sense, of course, only when you have reached the Moon.

第 8 章　语篇的翻译

8.1　语篇与语篇特征

常见的科技语篇主要包括科技著作、科技论文和报告、实验报告和方案；科技情报和文字资料；科技实用手册和操作规程；科技问题的会谈、会议、交谈的文字资料；科技的影片、录像、光盘等有声资料的解说词以及描写和解释大自然现象的语篇等。

科技语篇中句子结构大多环环相扣、逻辑紧凑。英语重形合，注重显性接应，注重句子形式，注重结构完整，注重以形显义；汉语则注重逻辑事理顺序，注重功能、意义，注重以神统形。英语的关系代词和关系副词使英语句子伸缩自如，形成“多枝共干”的“树形结构”，句子结构复杂而且较长。这些差异与特征也体现在科技语篇中。

例 1　A memory is a medium or device capable of storing one or more bits of information. In binary systems，a bit is stored as one of two possible states，representing either a 1 or a 0. A flip-flop is an example of a 1-bit memory，and a magnetic tape，along with the appropriate transport mechanism and read/write circuitry represents the other extreme of a large memory with an over-billion-bit capacity.

Computer memory can be divided into two sections. The section common to all computers is the main memory. A second section，called the file or secondary memory，is often present to store large amounts of information if needed. The main memory is composed of semiconductor devices and operates at much higher speeds than does the file memory. Typically a word or set of data can be stored or retrieved in a fraction of a microsecond from the main memory. We shall limit our discussion to semiconductor main memory. There are two broad classifications within semiconductor memories，the read-only memory（ROM）and the read-write memory（RWM）. The latter is also called a RAM to indicate that this is a random-access memory.

【译文】存储器是能够存储一位或多位信息的媒体或装置。在二进制系统中，一位以两种可能状态之一进行存储，分别代表 1 或 0。触发器就是一位存储器的例子。配有合适的传送装置和读写电路的磁带是大存储器的另一个极端的例子，存储能力在 10 亿位以上。

计算机的存储器可以分成两部分。所有计算机都有的部分是主存储器。第二部分称为文件存储器或辅助存储器，在需要的时候，常用以存储大量的信息。主存储器是由半导体器件组成的，其运行速度比文件存储器快得多。一般说来，以零点几微秒的时间即可对主存储器存或取一个字或一组数据。我们只讨论半导体型主存储器。半导体型主存储器分为两大类：只读存储器（ROM）和读写存储器（RWM）。后者也称为 RAM——随机存取存储器。

这个例子中，原文主要采用述位派生型的推进模式，使句子的意思环环相套。这种推进模式在中英文科技语篇中出现的比例较高，这种现象和科技文体本身的特点有关，也是科技英语描述客观世界经常采用的基本思维模式。

在科技语篇中，使用省略 by 短语的被动句可以增强科技英语描述的客观性和可信度。而英语被动语态具有突出主题和语篇衔接与连贯的功能。如果机械地翻译，保存原文主语的话，则不符合汉语的表达习惯。原文的第十句是英语的典型存在句，there be 结构是倒装结构，具有强调、衔接功能。所以译文按照汉语的习惯把 semiconductor memories 这个主题提前，后面部分作为评论。有了语篇意识的介入，翻译时必须要透彻地理解原文的语篇层面，在宏观的语篇框架下把单词、小句看成有机的一个整体进行理解。

科技文本的语篇特征包括两种：结构性的语篇特征和非结构性的语篇特征。前者指语篇内句子本身的结构，后者指话语内部的上下衔接。

8.2 结构性的语篇特征及翻译

语篇的结构性特征指的是语篇内句子的主位结构和信息结构。主位结构包括主位述位。从主位结构来看，一个结构完整的句子都由主位和述位构成，主位是全句的内容起点或谈论的话题，述位是对主位做出的描述，主位在前，述位在后。信息结构指的是已知信息与新信息相互作用而形成信息单位的结构。由于主位推进模式是组织语篇的重要语法手段，也是分析语篇组织的重要方法。因此，下面主要从主位同一型和延续型两种模式分析英汉科技文本的结构性语篇特征及翻译。

8.2.1 主位同一型

主位同一型也称平行型或放射型，指在一个语篇中以第一句的主位为出发点，以后各句以此句为主位，分别引出不同的述位，从不同的角度阐述这个主位。其典型特征是主位相同，述位不同：

T1——R1

T2（T2 = T1）——R2

T3（T3 = T1）——R3

（T 表示主位，R 表示述位，其右侧数字表示层次。）

例 2 The cables are normally made continuous through the tops of the towers, down through side towers, where these exist, and thence into the anchorage. They bear on specially constructed saddles on the towers, which are shaped to accommodate them, the saddles being either fixed so that the cables may slide over them, or mounted on rollers so that they move with any movement of the cables. In view of the enormous pull exerted by the heavy cables, their ends must be secured in firm anchorages, and unless they can be embedded in sound natural rock, constructions of masonry or concrete must be provided strong enough to withstand the severe pressures put upon them. The cable strands are normally looped round strand-shoes, which are in turn connected by chains to an anchor-plate embedded in the base of the anchorage.

【译文】通常，经塔墩顶部悬拉的缆索，要一直拉过两侧存在的桥墩，然后固定在嵌块中；缆索紧压在塔墩上面特制的滑座上。滑座根据缆索制成，可以固定，使缆索在上面滑动，也可安装在滚柱轴承上，使滑座随缆索任意移动。考虑到缆索很重，拉力较大，所以缆索端头必须牢固地固定在系缆嵌块上。如果缆索不能牢固地固定在天然岩石上，那么就应当构筑石材结构或混凝土结构，其强度应足以承受强大的作用力。通常缆索端头做成环套，然后通过铁链连接到系缆板上；而系缆板则灌注在嵌块的底座上。

在例2的四个复杂语句中，其主位都是 the cables（they、them、their ends、the cable strands），述位部分从不同角度对主位进行阐释，是典型的主位同一型的主位推进模式。将原文和译文进行比较发现，汉语尽量避免使用代词来阐述重复意义，而英语却往往借助于代词的照应或定冠词的应用。

8.2.2 延续型

延续型推进模式也称梯形模式，其信息分配原则是已知信息出现在句子前部，新信息出现在句子后部，而上句初步提到的新信息即是下句需要深化的已知信息，在深化时又引出新信息，如此环环紧扣、逐步完善。其典型特征是前一句的述位或述位的一部分成为后一句的主位，这种模式在科技语篇中很常见。如：

T1——R1

T2（T2 = R1）——R2

T3（T3 = R2）——R3

例3 Large quantities of steam are used by modern industry in the generation of power. It is therefore necessary to design boilers which will produce high-pressure steam as efficiently as possible. Modern boilers are frequently very large, and are sometimes capable of generating 300,000 lb of steam per hour. To achieve this rate of steam production, the boilers should operate at very high temperatures. In some boilers, temperatures of over 1650 °C may be attained. The fuels which are burned in the furnace are selected for their high calorific value, and give the maximum amount of heat. They are often pulverized by crushers outside the furnace and forced in under pressure.

【译文】在动力的生产方面，现代化工业使用了大量蒸汽。为此，需要设计出高效率产生高压蒸汽的锅炉。现代锅炉的生产能力往往都很大，有的一小时甚至能产生30万磅蒸汽。要达到这样的生产速率，锅炉就应在非常高的温度下工作。在某些锅炉中温度可达1650 °C。要从发热值高的角度出发选择炉内燃烧的燃料，就能发出最大的热量，燃料在炉外要先用粉碎机碾成粉状，然后在压力下压送进炉腔。

原文是由七个句子构成的延续型推进模式的语篇，述位变主位的话语结构明显。译文遵循了原文的推进模式，信息传递有着跟原文同样的“由已知到未知再到已知”的信息渐进结构，忠实传译了原文话语结构所指意义。

8.3 非结构性的语篇特征及翻译

英汉科技文本非结构性语篇特征指英汉科技文本不同句子中出现的不同成分之间的衔接关系，其表现形式有词汇衔接和语法衔接两种。语篇的衔接是语篇中的谋篇意义，是一种语义上的联系。如果篇章中的某一部分对另一部分的理解起关键作用，那么这两部分之间就存在衔接关系。

衔接理论在科技英语翻译中起着重要的作用。在语篇中，衔接点体现了句子间和段落间的联系，并且把句子和段落结为一个连贯的整体。因此，为了使译文清楚、简明地陈述事实、描绘客观现象，在翻译时需要运用衔接理论来理解原文，灵活处理原语和译语在衔接方面的差异，运用恰当的翻译策略来传达原文的意义，创造出一个符合译语使用规范的译文。

8.3.1 词汇衔接

词汇衔接指的是词汇的选用在语篇中所起的形成关联的作用。衔接在很大程度上是词汇关系而非语法的产物，词汇衔接是创造篇章结构的主要手段，占篇章衔接纽带中的40%左右。词汇衔接为篇章的连贯提供了基础，语篇的主题是由词汇衔接的贯通而得到实现的。

英语的词汇衔接关系被分为两大类：复现关系和同现关系。词汇的复现关系包括：重复、同义词或近义词、上义词和泛指词几种形式。词汇的复现关系是通过某一词以原词、同义词、近义词、上义词、下义词、概括词或其他形式重复出现在语篇中，语篇中的句子通过这种复现关系达到了相互衔接。同现关系指的是词汇共同出现的倾向性。具体翻译中可采用重复、增译和省略等方法。

8.3.1.1 重复

例 4 Meats sterilized by irradiation have much the same flavor, texture, color and nutritional value as their fresh counterparts.

【译文】经过辐射消毒的肉类，其色、香、味和营养价值与鲜肉相差无几。

例 4 中含有 meats 和 counterparts 两个异形近义词，构成了词汇衔接中的同义词或近义词关系（也是复现关系之一），同时这两个词本身有完全对等的汉语翻译“肉”与之对应，所以该译文中的衔接方式与原文保持了一致，即复现了原语的衔接手段。

例 5 Associated with limit loads is the proof factor, which is selected to ensure that if a limit load is applied to a structure the result will not be detrimental to the functioning of the aircraft, and the ultimate factor, which is intended to provide for the possibility of variations in structural strength.

【译文】与极限荷载有关的是保险系数和终极系数，保险系数是用来保证即使极限荷载加在飞机上，飞行也不会受到损害；终极系数则是用来预防结构强度可能发生的变化。

例 5 中 limit load 分别以复数和单数重复出现两次，在翻译的时候不考虑单复数的变化，均译为“极限荷载”。

例 6 In general drying a solid means the removal of relatively small amounts of water or other liquid from the solid material to reduce the content of residual liquid to an acceptably low value.

【译文】一般来讲，干燥一种固体指的是从固体材料中除去相对少量的水或其他液体，从而使残留液体的含量减少到可接受的低值。

原句中 solid 和 liquid 各出现两次，因此，译文中重复翻译了“固体”和“液体”这两个词汇，复现了原句的衔接方式。

8.3.1.2 增译

增译法是科技英语翻译中经常使用的翻译手段之一。在翻译中，应根据汉语的表达方式和习惯对原文进行增译，即在原文的基础上添加必要的单词、词组、分句或完整句，从而使译文在语法、语言形式上符合汉语的表达习惯并使译文在修辞、语法结构、词义或语气上与原文保持一致。

例 7 More than 2,000 patients are dying annually while waiting for transplants... The shortage of organs is so acute.

【译文】每年有两千多个等待器官移植的病人濒于死亡……（人体）器官的短缺非常严重。

在翻译 transplants 时，应在其前面增加“器官”一词，以达到汉语译文通畅及连贯。

例 8 A new kind of computer—cheap, small, light is attracting increasing attention.

【译文】一种新型电脑正在越来越引起人们的注意，它造价低、体积小、重量轻。

在翻译原文中的形容词 cheap、small、light 时，增加“造价、体积、重量”等词汇将之紧密衔接起来，使读者能理解原文的意思。

例 9 The machine is designed in the effort to minimize weight, dimension and manning.

【译文】设计这种机器的目的是最大限度地减轻重量，缩小体积，减少人力。

如果单根据字面意思，就会把原句后半部分译成“减轻重量、体积和人力”，但是这样翻译不符合汉语的表达习惯，因为“减轻”可以与“重量”搭配，但是与“体积”和“人力”搭配就不太贴切。因此，在翻译过程中应该分别增加“缩小”、“减少”与“体积”和“人力”搭配，使表达更加准确。

8.3.1.3 省略

例 10 Semi-conductors have a lesser conducting capacity than metals.

【译文】半导体的导电能力显然比金属差。

原文中的动词 have 在翻译中可以省略，使句子前后语义更衔接，使译文更加符合汉语表达习惯。

例 11 Some robots can find the necessary parts and assemble them according to drawing using a television eye and a mechanical hand and arm.

【译文】某些机器人可以用电视摄像机和机械手找到所需的零件，并按图纸把它们装配起来。

有些同义名词在英语中常连用，以表示同一名称的两种不同说法。由于意义相似，所以句中的 hand 翻译出来，但 arm 就省译了。

8.3.2 语法衔接

语法衔接是实现语篇连贯的多种衔接机制之一。语法衔接是指利用语法手段使句际和句组之间达到上下文衔接的目的。在科技英语翻译中，由于英汉两种语言在词法、句法和衔接方式上均存在差异，其思维方式和表达方式也不尽相同，这就决定了原语和译语之间不能完全进行等值翻译。因此，翻译时要使译文合乎汉语的习惯和表达规律，使意思更加明确清晰，使译文语义完整、语言表达自然流畅，就需要采取相应的翻译策略。

8.3.2.1 照应

照应是语篇中的指代成分与指称或所指对象之间的相互解释关系，是语篇实现其语法上，从而结构上的衔接和语义上的连贯的一种主要手段。

在语篇中，如果对于一个词语的解释不能从词语本身获得，而必须从该词语所指的对象中寻找答案，这就产生了照应关系。照应作为语篇的语法衔接手段之一，通过恰当地使用人称代词、指示代词、副词以及表示比较意义的限定词等使语篇语义连贯，层次分明，实现篇章结构的严密性和连贯性。英语中的照应系统分为：人称照应、指示照应、比较照应。

（1）人称照应

人称照应是指用人称代词及其相应的限定词和名词所有格代词所表示的照应关系。因此，在人称照应系统中只有第三人称代词具有内在的语篇衔接功能，因为第三人称代词主要用在回指中，而第一、第二人称代词所指代的分别是发话者和受话者，所以一般不具有语篇衔接功能。人称照应即用代词复指上文（回指）或预指下文（下指）出现的名词的一种衔接手段。人称照应在汉英语篇中有差异。首先，汉语中没有关系代词，而英语区别于汉语的一个重要特征就在于有关系代词。所以在很多情况下，汉语语篇中的人称代词在对应的英语表达中可以用关系代词表示。

例 12　Vitamin E is necessary to prevent sterility. If it is absent from the diet, female animals are unable to complete pregnancy successfully since the embryo die and are absorbed.

【译文】维生素 E 是预防不育症所必须的。膳食中少了它，雌性动物不能成功完成妊娠，因为胚胎将会死亡并被吸收。

原句中的人称照应词 it 在上文中能找到明确的指代对象 Vitamin E，翻译成汉语时可保留其衔接手段，用“它”来回指“维生素 E”。

例 13　Electrical inventors who followed Edison did not have to experiment with the substances which he had found would not work.

【译文】爱迪生发现某些材料无效，这使继他之后的电器发明家们无须再对这些材料进行试验了。

原句中的人称照应词 he 在上文中的指代对象是 Edison，翻译时利用重复法重复其所代替的名字。

（2）指示照应

指示照应是指用指示代词或相应的限定词以及冠词等所表示的照应关系。在指示照应中，时间和空间上的远近是以发话者所在的时间和位置作为参照点的。就指示词所指时间和空间概念而言，this、these、now 和 here 指近，that、those、then 和 there 指远，而 the 是中性的。除此之外，还有单复数之分，即 this 和 that 为单数，there 和 those 为复数。当用做中心词时，this 和 that 常用于回指上文所说的内容。当回指自己所说的话时一般用 this，而回指对方所说的话时一般用 that。

例 14　When regeneration is not needed, it is possible to use the half-controlled bridge shown in Fig 8-2. This circuit has thrusters on two arms. It produces full-wave rectification but has one major disadvantage in that it is essential to ensure thruster is extinguished before the start of its positive half-cycle of the system voltage.

【译文】如果无须再生电流，还可使用图 8-2 的半控制电桥。该电路两臂上有硅控整流器，其阴极与正极和另两臂上的整流器相连。它能产生全波整流，但也有明显的缺点，即务必确保系统电压正半周期启动前各硅控整流管为封闭状态。

原句中 this 和 it 都回指 regeneration circuit，直接翻译成对应的代词。汉译时，若重复代词所指的词时就会显得多余，而若省译则会使语篇不连贯，所以最好保留相应的代词。

此方法较常用于指示照应和比较照应的翻译。

例 15 The properties of alloys are much better than those of pure metals.

【译文】合金的性能比纯金属的性能要好得多。

原句中 those 回指 properties。在汉译时直接翻译成所指代的名词“性能”。

（3）比较照应

比较照应是指用比较事物异同的形容词或副词及其比较级所表示的照应关系。因为任何比较都至少包含两个实体或事态，所以当语篇中出现表达比较的词语时，受话者就会从上下文寻找与之构成比较关系的词语。在汉语中常见的形容词有“同样的”、“相同的”、“同等的”、“类似的”、“不同的”，还可以运用一些比较结构来表达一般比较，常用的有“如……一样”、“和……差不多”、“和……相同”、“像……之类的”、“不像……那样”、“……也是如此”等。总之，英汉两种语言的比较照应差异不大，主要都借助形容词和副词来表达比较意义。但是，句内衔接说明句子结构本身具有良好的衔接性，所以其研究意义不如句际衔接的研究意义重大。

例 16 The embryo has been implanted in his mother and when the baby is born it will be used to give his brother a blood transfusion that could save his life. Another woman is also pregnant in a similar procedure in the hope of curing her child of an inherited blood disease.

【译文】这个胚胎已植入母体，当婴儿出生后，将被用来给他的哥哥进行输血，以拯救他的生命。另一名妇女也用同样的方式受孕，以期治疗她那患有遗传性血液病的孩子。

原句中的 a similar procedure 是典型的比较照应，说明两者所经历过程的相似性。在汉语中为实现信息的对等，应保留原文中的比较照应关系，可用“同样的”、“类似的”等表达形式来翻译。

8.3.2.2 替代

为了避免重复，表达简洁，结构紧凑，但又不导致歧义，常用语法衔接手段替代，即用 one、do、so、thus、ones、the same 等替代上文提到而又在下文出现的成分。翻译时除省略情形外，为避免指代混乱，有时要重复代词所指的名词。

例 17 Just as man has found great uses for the materials which he can dig up from the ground, he has found important uses for the gases which he can obtain from the air.

【译文】正如人类发现从地下采掘出来的物质有巨大的用途一样，人类也发现了从空气中取得的各种气体亦具有重要的用途。

原文中用人称代词 he 复指了上文提到的 man，翻译时则重复指代名词 man，译成“人类”。

例 18 There are two fire-boxes inside the boiler, an inner one and an outer one, which extend a long forward.

【译文】锅炉有两个火箱，一个内火箱，一个外火箱。这两个火箱都向前延伸很长一段距离。

原句中 fire-boxes 在后面用两个 one 替代了，但翻译成汉语时将名词重复译出，更符合汉语习惯。

例 19 An electric light bulb is a vacuum, and so is a radio tube.

【译文】电灯泡是真空的，电子管也是真空的。

原句中so代替主句中的vacuum，汉译时要明确指出替代的名词，使语义更加明确。

例20 It is clear from the figure that senior managers need both the past and future information related to equipment performance, the shop floor level personnel need the same.

【译文】从图中可清楚看出高层次管理人员需要与设备性能有关的过去和未来的信息。基层修理车间的人员需要同样的信息。

原句中the same使用了替代，译文加入"信息"二字，符合汉语表达习惯，并实现其语义对等。

8.3.2.3 省略

省略法是重要的语篇衔接手段之一，是与增译法相对的一种翻译策略。省略也是一种特殊的替代形式，通常被称为"零替代"。翻译时可在不改变原文意思的基础上省略上下文已提到的交际双方可以填补的不必要的成分。省略的使用可避免行文重复，而且使上下句之间在语义上建立起一种互相依赖的衔接关系。

例21 If it is hollow, a ship can displace a sufficient volume of water to buoy it up.

【译文】船是空心的，能排开足够的水浮起来。

原句中的连词if和代词it在译文中均被省译。

例22 Different metals differ in their electrical conductivity.

【译文】不同金属具有不同的导电性能。

例23 Hydrogen is the lightest element with an atomic weight of 1.0008.

【译文】氢是最轻的元素，原子量为1.0008。

例22中的their可以省略不译；例23原句中的介词with在译文中省译。

8.3.2.4 增译

运用增译法对科技英语的译文进行处理，是科技英语翻译中经常使用的翻译手段之一。

例24 They've met with software and/or hardware problem(s).

【译文】他们碰到过软件问题，也碰到过硬件问题，还碰到过软件和硬件都出问题的情况。

由and/or连接的句子在技术指南和产品说明书中经常运用。译者应掌握这种结构的含义。翻译时不能译成"他们碰到过软件问题，也碰到过硬件问题"，而应该从结构上增译成"还碰到过软件和硬件都出问题的情况"。

8.3.2.5 连接

连接是通过连接成分体现语篇中句与句之间的各种逻辑联系。连接成分通常分为四种类型，即增补、转折、因果和时间。但句子间的逻辑关系有两种情况：有各种连接词，如but、however、therefore、consequently等；没有连接词。在后一种情况下，必须探索各句子在语义上的内在逻辑关系，才能得出较好的译文。

例25 The exact composition of the corrosion product depends upon that atmosphere so that the patina may be protected or permit further corrosion leading to deep and penetrating attack.

【译文】腐蚀产物的确切成分取决于环境因素，所以铜绿可能会保护金属，也可能会让金属受到进一步的腐蚀，进而向深部渗透。

例25中，主句和从句之间由表示因果的连接词so that连接，翻译的时候直接翻译成带因果的主从句即可。

例 26 The ever-increasing world-wide demand over the last decade for telecommunication facilities of all kinds has resulted in dramatic developments both nationally and internationally. Advances in electronic techniques and the application of new materials and devices resulting from scientific research lie at the heart of the enormous achievements that have been made.

【译文】近十年来全世界对于各种电信设备的需求日益增长，使得国际、国内电讯事业迅猛发展。这些巨大成就的取得主要是由于科学研究促使电子技术的发展和新材料设备的应用。

原句如果按其原来顺序生硬翻译成“近十年来全世界对于各种电信设备的需求日益增长，使得国际、国内电讯事业迅猛发展。科学研究促使电子技术的发展和新材料设备的应用是取得这些巨大成就的主要原因”，就会使两个句子在意思上不连贯，逻辑上衔接不紧密。如果将第二个句子颠倒译为：“这些巨大成就的取得主要是由于科学研究促使电子技术的发展和新材料设备的应用”，则不仅可以与第一句很好地连接起来，而且还突出了这一段的主题“电讯事业的迅猛发展”。

例 27 Until the maglev train is moving fast enough to lift off, it rolls on wheels that retract as soon as the maglev hits 106 mph.

【译文】磁浮列车先是靠车轮滚动前进，达到一定时速就升离地面，而当其时速达到 106 英里，车轮就收缩起来。

这句话是描写磁浮列车的运行步骤的，按事情发生的逻辑顺序来翻译，可以使译文连贯流畅，因此需要打破原句结构的制约，重组译文。

翻译练习

1. The concept that specific immune responses serve to enhance natural immunity is also reflected in the phylogeny of defense mechanisms.

2. Typhoon and hurricanes do not form exceptionally large waves, because their winds, although very strong, do not blow long enough from one direction.

3. It is clear from the figure that senior managers need both the past and future information related to equipment performance, the shop floor level personnel need the same.

4. Pressure pulsations have little influence on the seal effectiveness and reliability of the joint, as long as the compression of the elastomeric ring is not eliminated or in any other way affected.

5. The volume of the sun is about 1,300,000 times that of the earth.

6. The laws in science are frequently stated in words, but more often in the form of equations.

7. Any substance is made up of atoms whether it is a solid, or a liquid, or a gas.

8. Traditionally, chemistry has evolved onto four provinces: organic, inorganic, physical and analytical chemistry.

9. The world needn't be afraid of a possible shortage of coal, oil, natural gas or other sources of fuel for the future.

10. Automation involves a detailed and continuous knowledge of the functioning of the system, so that the best corrective actions can be applied immediately they become necessary. Automation in this true sense is brought to full fruition only through a thorough exploitation of its three major elements, communication, computation, and

control——the three 'Cs'. I believe there is a great need to make sure that some, at any rate of the implications to our society of three 'Cs' in combination are recognized and understood.

11. CO is a colorless, odorless, tasteless gas that fortunately is not known to have adverse effects on vegetation, visibilities, or material objects. Its dominant impact appears to be toxicant to man and animals, which arises from its well-known competition with hemoglobin in red blood cells, as in equation 3. 1.

12. By Aristotle's time, people had for centuries been recording how the lights in the night sky moved.

13. The pressure of increasing population leads to the vertical growth of cities with the result that people are forced to adjust themselves to congestion in order to maintain these relatively artificial land values.

14. All known particles carry an electric charge which is exactly equal to or a multiple of the electron charge.

15. The downward pressure exerted by water is proportional to its depth.

第 9 章　标点符号的翻译

正确地使用标点符号，能够准确、有效地表情达意。字词相同，标点相异，表达的意思则大相径庭。由于英汉标点符号的使用规范和习惯有所不同，英译汉时也必须适当地加以变通，否则就可能曲解原意。

9.1　英汉标点符号

根据中国语言文字工作委员会和新闻出版署发布的《标点符号用法》，汉语书面语共有 16 种标点符号。标点可分为点号和标号。点号含句号、问号、感叹号等句末点号和分号、冒号、逗号、顿号等句中点号两类。标号主要指引号、括号、书名号、专名号、破折号、间隔号、连接号和着重号等。综合多种相关著作和实践运用的情况可知，英语也有 16 种标点符号，而且与汉语标点符号多数可以对应或转换。表 9－1 是英汉标点符号对照表。

表 9－1　英汉标点符号对照表

标点符号	英语符号	汉语符号
句号	.	。
问号	?	？
感叹号	!	！
分号	;	；
冒号	:	：
逗号	,	，
顿号	无	、
引号	“” 或 ‘’	“” 或 ‘’
括号	() 或 [ ]	（） 或【 】
省略号	...	……
破折号	——	——
间隔号	无	·
连接号	-	—
着重号	异体	.
书名号	斜体	《 》
专名号	无	_ _ _ _
连字号	-	无
斜线号	/	无
撇号	'	无

有些英语的标点符号在翻译时可以直接转换成对应的汉语的标点符号。

例 1　This is a college of science and technology, the students of which are trained to be engineers or scientists.

【译文】这是一所科技大学，该校学生将被培养成工程师或科学工作者。

例2　That car was really moving!

【译文】那辆汽车真快!

例3　The Theory of Relativity worked out by Einstein is now above many people's comprehension.

【译文】爱因斯坦提出的“相对论”现在还有许多人理解不了。

例1中的逗号、例2中的感叹号及例3中的句号都在译文中分别转换成汉语的对应符号。

9.2　英语标点符号的转换

由于英汉标点符号不是一一对等的，而且在翻译时英汉语言结构也必然发生不同程度上的变化，因此对标点符号也必须做出相应的变通。

9.2.1　逗号

英语中的逗号一般用在句首的状语从句之后，并列的分句之间，同位语或者插入语的前后。汉语的逗号主要用于主语之后，句首的状语之后，并列词组之间，独立语之前或者之后以及关联词后。在英译汉时，要根据句子表达的意义对逗号进行转换。

9.2.1.1　逗号转换为顿号

英语中逗号的许多用法和汉语相当，但它在连接一个句子中的并列成分时，多转换成汉语的顿号。

例4　All the other metals then known, such as tin, lead, copper, aluminium and iron, were by contrast considered “base metals”.

【译文】相比之下，当时已熟悉的其他金属，如锡、铅、铜、铝和铁，便看做是“贱金属”了。

例5　This substace makes up carbon, hydrogen, oxygen and sulfure.

【译文】这种物质是由碳、氢、氧和硫组成的。

例4中第2、3、4个逗号和例5中的两个逗号都在译文中转换成了顿号。

例6　He succeeded by accident when he overheated a mixture of rubber, sulphur, carbon and white lead.

【译文】当他把橡胶、硫、碳和铅白混合在一起过度加热时，他碰巧获得了成功。

9.2.1.2　逗号转换为冒号

英语中的“某某说”之类的词语在直接引语之前，其后多用逗号，在汉语中则应改用冒号。

例7　He said, “Necessity is the mother of invention.”

【译文】他说：“需要是发明之母。”

例8　The teacher told the students, “Columbus discovered America in 1842.”

【译文】老师告诉学生说：“哥伦布在1842年发现了美洲。”

9.2.1.3　增加逗号

英译汉时，有时要根据具体情况增添原文中没有的逗号，使译文既忠实于原文，又符合汉语的习惯。

例 9　Energy can neither be created nor destroyed although its form can be changed.

【译文】能量既不能创造，也不能消灭，尽管其形式可以转换。

原句中只有句末一个句号，但译文增加了两个逗号。

例 10　It is man who plays the leading role in the application of electronic computers.

【译文】在使用电子计算机时，起主要作用的是人。

原句是强调句，翻译时增加了逗号。

在翻译一些状语从句时，可增添逗号与主句隔开，使表述更加清楚。

例 11　Heat can flow from a hot body to a cooler one as if it were a fluid.

【译文】热能从一个热的物体传到一个较凉的物体上，好像流体一样。

例 12　The material first used was copper for the reason that it is easily obtained in its pure state.

【译文】最先使用的材料是铜，因为纯铜易于制取。

例 13　This kind of plane cannot be built unless we find a metal even lighter than this high-strength aluminum alloy.

【译文】我们不可能造出这种飞机来，除非我们找到一种比这种高强度的铝合金更轻的金属。

例 14　All metals will melt though some require greater heat than others.

【译文】所有金属都会熔化，虽然有一些金属熔化时比另一些金属需要更高的温度。

例 11 ~14 分别是方式状语从句、原因状语从句、条件状语从句和让步状语从句，从句前均没有逗号，但译成汉语时用逗号将从句与主句分隔。

例 15　The vapor pressure of water increases as the temperature is raised.

【译文】温度升高，水的蒸汽压也增高。

原句中状语从句较短，也可以增添逗号转译成并列分句。

例 16　It is obvious that oil is lighter than water.

【译文】显然，油比水轻。

例 17　Scientists have proved it to be true that the heat we get from coal and oil comes originally from the sun.

【译文】科学家已证实，我们从煤和石油中得到的热都来源于太阳。

例 16、例 17 是以 it 作形式主语（宾语）的名词性从句，带有强调的意味，翻译时可增添逗号与主句隔开。

例 18　Engineering metals are used in industry in the form of alloys except that aluminum may be used in the form of a simple metal.

【译文】除了铝可以纯金属形态使用外，各种工程金属都是以合金形式应用于工业。

英文中以 except、but 等词引导的宾语从句放在句末，但按汉语习惯一般应倒置在主句之前。因此，例 18 增添逗号与主句隔开。

例 19　He spoke with understandable pride of the invention of the instrument.

【译文】他谈到那种仪器的发明时很自豪，这一点是可以理解的。

英语中的有些形容词在译成汉语时经常翻译成一个小分句，这个小分句的前面或者后面需添加逗号。

例 20　Two of the advantages of the transistor are its being small in size and its being able to be put close to each other without overheating.

【译文】晶体管有两个优点，一是体积小，二是相互能靠近放置而不过热。

原句在译成汉语时把原来的简单句进行了拆分，添加了两个逗号，以符合汉语的表达方式。

9.2.1.4　省略逗号

翻译时，英语句子中的插入语可按照汉语的表达习惯译成含有前置定语“……的”句型的简单句，而且，出于汉语结构上的需要，插入语前后的逗号应省略。

例 21　Transistors，which are small in size，can make previously large and bulky radio light and small.

【译文】体积小的晶体管使先前那种大而笨的收音机变得又轻又小。

原句中的非限定性定语从句 which are small in size 前后的逗号都可省略，将原从句改为前置定语，译成“……的”的句型。

例 22　Hydrogen burns in air or oxygen，forming water.

【译文】氢气在空气或氧气中燃烧便形成水。

例 23　In an absorption system，the refrigerant is usually ammonia.

【译文】吸收系统通常用氨作冷却剂。

例 22、例 23 的译文中省略了原句中的逗号。

9.2.2　句号

在英语和汉语中，句号都是使用频率最高的标点符号。然而，英语和汉语在句号的使用上还是有所差异的。

9.2.2.1　句号转换成逗号

汉语在句中多用逗号，但英语在句中有时用句号独立成句。因此，翻译时可根据汉语表达习惯将英语的句号改为汉语的逗号。

例 24　There will be an experiment to carry out. Are you going to take part in?

【译文】要进行试验了，你参加吗?

例 25　Aqueous solution of certain compounds will conduct an electric current. Such compounds are called electrolytes.

【译文】一些化合物的水溶液导电，这样的化合物叫电解质。

例 24、例 25 在汉译时均将句中的句号改成了汉语的逗号。

9.2.2.2　增加句号

汉语一般不用长句，因此在英译汉时，往往要把一个不易安排的单词或短语译成一个分句，或把一个较长的句子拆开译成两个或两个以上的句子，以使译文层次分明，合乎汉语习惯，并对标点符号做出相应的变通，增添必要的句号。

例 26　Flying is the oldest dream of man，perhaps reaching back to the days when caveman looked longingly upward at the birds flying overhead as they wearily pursued the animals on the earth beneath.

【译文】飞行是人类最古老的梦想，这也许要追溯到穴居人时代。那时侯，当他们在地面上追捕动物累得筋疲力尽时，仰头看着在头顶上飞翔的鸟儿，无限向往。

9.2.3 冒号

9.2.3.1 增加冒号

英文中不使用冒号，但在译文可考虑添加冒号，使汉语表述更准确顺畅。

例 27 Always, though, there is the underlying idea that organized sport is a valuable and productive use of a young person's time.

【译文】然而，总是有一个基本想法：有组织的体育活动是对青少年时间的很有价值、很有成果的利用。

原句在列举的事物、引用的句子前有 there be 时，译文可在这些事物、句子前加上冒号。

9.2.3.2 转换为冒号

例 28 As so often happens, the darkest hour comes just before the dawn.

【译文】事情往往是这样：黎明之前最黑暗。

原句中使用了逗号，逗号后是解释性分句。因此，汉译时可将逗号转换成冒号。

例 29 And there was possibility that a small electrical spark might accidentally bypass the most carefully planned circuit.

【译文】而且总有这种可能性：一个小小的电火花，可能会意外地绕过最为精心设计的线路。

原句是同位语从句，出于汉语的表达习惯，常在相关词后增添冒号，再将从句顺译于后。

9.2.4 破折号

9.2.4.1 省略破折号

中英文中均有破折号，破折号表示它后面有个注解、附加成分，或者表示语意的转折、递进，它是正文的一部分。汉译英时可视情况突破原文的文字结构和顺序，对标点作相应的调整。

例 30 Other living forms have become highly complex——insects, vertebrates, ferns, conifers and flowing plants.

【译文】有些生物，例如昆虫、蜘蛛、脊椎动物、蕨类植物、针叶树和显花植物，已经变得非常复杂。

原句中的破折号（在原文中起注解作用）是因为主语的同位语被谓语分隔而加的，而中文不允许这样的分隔，因此，翻译时可省掉。

9.2.4.2 增添破折号

汉译英时，为有效地进行注释、补充说明，译文中有时要增添破折号。

例 31 If you go to visit Nobel's old residence, the house in which the great chemist remained a bachelor throughout his life, you will catch sight of a shelf laden with experimental records.

【译文】如果你参观诺贝尔的故居 —— 在那座房子里，这位伟大的化学家过了一辈子的独身生活 —— 你将会看到一个堆满试验记录的书架。

例 32 A few years later he was able to take out his first patent for transformers.

【译文】数年之后，他获得了发明变压器的专利 —— 这是他获得的第一项专利。

9.2.4.3 转换成括号

英语的破折号引出的是对前面相关词的进一步解释说明时，汉语译文中可以相应地转换成括号形式。如：

例 33 We'll live longer (120 years?) . If the normal aging process is basically a furious, invisible contest in our cells——a contest between damage to our DNA and our cells' ability to repair that damage —— then 21st century strides in genetic medicine may let us control and even reverse the process.

【译文】我们将活得更长（120 岁?）。如果说通常的衰老过程主要是我们细胞内的一场激烈而不可见的竞赛的话（一场对我们的脱氧核糖核酸进行破坏同我们体内的细胞对破坏的组织进行修复的竞赛），那么 21 世纪遗传医学的巨大成就会使我们能够控制，甚至逆转这一过程。

9.2.5 分号

英语中的分号主要用于连接在语法上互不依赖、意思上相互关联但不一定是并列关系的两个或两个以上的分句。汉语中的分号主要用于从结构和意思上均并列的几个分句之间。由此可知，英语中分号的应用比汉语广泛。所以，英语中一些以分号连接的结构简单的并列复合句，译成汉语时大多要改用逗号连接。

例 34 Water and air are both necessary to man; the later is more important.

【译文】水和空气都是人类所必须的，后者更为重要。

例 35 The task is difficult; moreover, time presses.

【译文】任务艰巨，时间紧迫。

例 36 Sulfure is pale yellow; bromine is reddish brown; chlorine is greenish yellow.

【译文】硫是淡黄色，溴是红综色，氯是黄绿色。

9.2.6 引号

直接引用现成语句，英汉语句末标点均可置于引号之内；引述部分文字时，英语既可将句号置于引号之内，也可将句号置于引号之外，但汉语则通常将句号或由于译文结构变通而成的逗号置于引号之外。

例 37 "Why table salt dissolves in water and diamonds do not?" asked Tom to his father.

【译文】"为什么食盐溶于水而金刚石不溶于水呢?" 汤姆向他父亲问道。

例 38 That is why practice is the criterion of truth and why "the stand point of life, of practice, should be first and fundamental in the theory of knowledge".

【译文】所谓实践是真理的标准，所谓"生活、实践的观点，应该是认识论的首要的和基本的观点”，理由就在于此。

例 39 He appealed to their “self-responsibility” asking them to “summon up the courage for fundamental changes together.”

【译文】他呼吁大家“自己负责”，要求大家“鼓起勇气共同去进行深刻的变革”。

9.2.7 斜杠

英文的斜杠多用做数字符号，如分数、日期等，也表示“每”、“或、”“和”。译成汉语时可用斜杠或者“每”。

例 40 The speed of sound in the air is 340 m/sec.

【译文】声音在空气中的传播速度是每秒 340 米。

例 41 The man could be sentenced to five years in jail and/or a $50,000 fine.

【译文】这个人可能被判处 5 年监禁 和（或）5 万美金的罚金。

翻 译 练 习

1. 10：00A. M. —5：00 P. M.
2. 1968—72
3. winter of 1944—45
4. Mediterranean islands and Canary Islands
5. At that time there were two important things which were worth mentioning：one was the practical knowledge of the Egyptian workers in metals，pottery and dyes；the other was the study of the earlier Greek philosophers.
6. Some Greek philosophers held that all matter was made up of the same four “elements” —— earth，fire，air and water.
7. If the transgenic animal is fertile，the inserted foreign gene（transgene）will be inherited by future progeny.
8. We'll have a brain road map. This is the real final frontier of the 21st century：the brain is the most complex system we know. It contains about 100 billion neurons（roughly the number of stars in the Milky Way），each connected to as many as 1,000 others early in the next century，we will use advanced forms of magnetic resonance imaging to produce detailed maps of the neurons in operation.
9. Best surface finish is provided by machining methods especially by grinding.
10. Hydrogen，which is the lightest element，has only one electron.
11. Electronic computers，which have many advantages，cannot carry out creative work and replace man.
12. This university has six faculties，namely，Computer Science，High Energy Physics，Laser，Geophysics，Remote Sensing and Genetic Engineering.
13. Mr. Smith，senior correspondent of The Times，London.
14. In 1872，he published the results in a paper titled the Galvanic Chain，Mathematically Treated.
15. ... great lorries with a double deck cargo of cars for export lumber past Magdalen and the University Church.
16. Within seven years，build according to specific requirements the different kinds of roads（highways，roads and paths）needed in the provinces，prefectures，counties，districts and townships.
17. In October 1998，Time magazine named the 50 most influential cyber space people in the world. Of them，

two are Chinese. One is Liu Yunjie, who is in charge of the data communication networks built by the Chinese government, and was called by the magazine "the father of China's Internet."

18. There is enough of a difference here to indicate that one must look at the fore-man's job in terms of what his situation is, whom he has to motivate and what opportunities he has to do——before deciding what sort of supervisor training is best for him.

19. Stainless steels possess good hardness and high strength.

20. For convenience of study, this body of knowledge is customarily divided into the classification: mechanics, heat, light, electricity and sound.

翻译练习参考答案

第1章

1. 如果没有改变物体运动的原因，那么物体将作匀速直线运动。

2. 电能可储存在由一绝缘介质隔开的两块金属极板内。这样的装置称之为电容器，其储存电能的能力称为电容。电容的测量单位是法拉。

3. 结构材料的选择应使其在外界条件中保持其弹性。

4. 这样的缓慢压缩能使这种气压经历一系列的状态，但各状态都很接近于平衡状态，所以叫作准静态过程，或“近似稳定”过程。

5. 计算机可分为模拟计算机和数字计算机两种。

6. 新型晶体管的开关时间缩短了2/3（或“缩短为1/3”）。

7. 尽管有上述5.6条（a）、（b）款的规定，转让方可将持有的合营公司注册资本份额部分或全部按下列条款转让给其某一关联机构（“关联受让方”）。

8. 具有突出防锈性能的不锈钢含铬的百分比很高。

9. 与数字式电视机相结合，图像光盘不仅可以演电影，还提供环境声音，产生电影院效果——令人吃惊的真实感，使观看者产生一种错觉，以为他们是在现场目睹他们周围发生的一切。

10. 这个中心正在举办关于科学与文学关系的系列讨论会。与会的诗人和科学家们讨论的议题是：在过去两个多世纪里科学思想对文学和社会变革的影响。

第2章

1. 因为物质的质量不变，所以增大作用力便会增加速度。

2. 改善城市空气质量最好的办法可能还是控制汽车的使用，尽管现代汽车比以前的汽车污染要小得多。

3. 自由电子往往做不规则运动。

4. 高温不但没有增加该反应速率，反而减低反应速率。

5. 这些水泵的特点是操作简单，维修方便，经久耐用。

6. 只有当火箭速度达到每小时18,000多英里时，才能把人造卫星送入轨道。

7. 在海湾的南部，从紧靠入口处内侧直到第二平潮区，其间距离大约是到海湾的一半，潮流正在回落。

8. 但是现在人们意识到，其中有些矿物质的蕴藏量是有限的，人们甚至还可以比较合理地估计出这些矿物质“可望存在多少年”，也就是说，经过若干年后，这些矿物的全部已知的矿源和储量将消耗殆尽。

9. 所用资料集为美国国家气象中心关于1962/1963到1981/1982期间20个冬季的一天两次的500百帕高度场分析资料。

10. 对链式反应必须谨慎控制。这要用硼控制棒吸收多余的中子来完成。热量由循环流动的二氧化碳气传送。高温气体则用于产生水蒸气。整个组合体称为核反应堆。

第3章

1. 沿节圆所得测得的齿厚是周节的一半。

2. 这次地震震级为6.5里氏震级。
3. 我们必须考虑一下如果床层意外崩塌应采取什么措施。
4. 在晶体管中，输出电流取决于输入电流，因此，晶体管是电流控制器件。
5. 继电器由数毫安电流启动。
6. 助听器用电池供电。
7. 在用碳还原氧化铁时，两种反应中的每一种都可能发生。
8. 合金是介于混合物和化合物之间的中间结构。
9. 钉子同时绕两个互相垂直的轴旋转。
10. 电子是绕原子核转动且带有负电荷的微粒。
11. 方向舵的用途是使飞机能够左右偏航。
12. 这几种化合物的沸点各不相同。
13. 经过一些试验后，伽利略成功地造出了一架好得多的望远镜。
14. 在两个半小时的商谈中，双方就付款方式交换了意见，但却没提到运输方式。
15. 使用电子计算机可以大大提高劳动生产率。
16. 那艘新建轮船的首航是成功的。
17. 地球的这一巨型保护伞由一层臭氧组成，其厚度约为20公里。
18. 我们仔细地研究了这些化学元素的特性。
19. 气体和液体具有很好的弹性。
20. 那个无线电厂给了我很深的印象。
21. 所有的奥氏体不锈钢都具有抗氢腐蚀性能。
22. 把一个磁石的北极对着另一个磁石的北极，这两个磁石就会互相排斥。
23. 这个实验虽然原理简单，却导致了一场影响深远的科学革命。
24. 一份列举柴油机所有用途的清单可能要好几页。
25. 因为空气中二氧化碳的浓度只有万分之三，所以聚集在植物内部的碳是从周围极大体积的空气中得来的。
26. 冰和水由相同的物质构成，但形态不同。
27. 假如没有重力，地球周围就不会有空气。
28. 科学家经常设法否定自己的假设，推翻自己的理论，并放弃自己的结论。
29. 固体传声效果很好。
30. 这部打字机真是价廉物美。
31. 詹姆斯·瓦特发明了蒸汽机。
32. 你给我时间，我能完成这工作。
33. 不带地图是会迷路的。
34. 我们应当逐步消灭城乡差别。
35. 求职者中，有工作经验者将优先录用。
36. 他双手插进口袋，然后耸了耸肩。
37. 可见光的波长范围约为0.38～0.78 μm。
38. 群众齐心了，一切事情都好办。
39. 从铁矿石中提炼铁不是那么容易的。
40. 现在争论是没用的。
41. 正是美国在歪曲颠倒“日内瓦精神”。
42. 为生存而吃饭，而不应该为吃饭而生存。
43. 工业废气对我们是有害的，应尽力排除。

44. 要掌握好一门外语，非下苦功不可。
45. 聪明的人不会犯这样的错误。
46. 地球围绕太阳转。
47. 广东省的人口比英国（的人口）多。
48. 如果知道频率，就能求出波长。
49. 干扰机的工作频段为 20 ~ 500 MHz。
50. 铝合金可分为两类：可热处理铝合金和不可热处理铝合金。
51. 一切金属都是良导体，因为金属里有大量的电子。
52. 爱迪生之后的电器发明家们无须再对爱迪生所发现的不管用的材料进行试验了。
53. 角焊缝所受外载基本上按两个方向作用：一个与焊缝平行，另一个与焊缝垂直。
54. 桁架可分为简单桁架、复合桁架和复杂桁架。
55. 功不包括时间，但功率包括时间。
56. 我们本想去实验室，但忘去了。
57. 我们必须分析问题，解决问题。
58. 市场经济本身是一个长期发展过程的产物，是生产方式和交换方式一系列变革的产物。
59. 在这四幅画中，有两幅看起来是真的，另外两幅看起来则是赝品。
60. 事实并非如此，因为这种问法是以人们对人的权利有共同认识为基础的，而这种共同认识并不存在。
61. 中国人民历来是勇于探索、勇于创造、勇于革命的。
62. 我们的政策，不光要使领导者知道，干部知道，还要使广大群众知道。
63. 幸福的家庭也有幸福家庭的苦恼。
64. 他既精于飞行，又善于导航。
65. 我们必须积极采用新技术，新设备、新工艺、新材料。
66. 我曾碰到过，不是氧气设备了故障，就是引擎出故障，或两者都出故障。
67. 在常压下，水在零摄氏度时变成冰，在一百摄氏度时变成蒸汽。
68. 我们首先需要的是：学习，学习，再学习。
69. 勇之过度则为蛮，爱之过度则为宠，俭之过度则为贪。
70. 我们必须明白科学是永无止境的。
71. 冰的密度比水小，因此能浮在水面上。
72. 金属要加热到一定程度才会融化。
73. 基因学家们近年来才开始对特定基因感兴趣。
74. 我们做实验时要尽可能小心。
75. 终端产品与实验产品有明显的差别，这种情况是常见的。
76. 只要温度不变，各种分子的平均速度也就不变。
77. 温度对金属的导电性影响不大。
78. 一般的非金属材料比金属材料的强度差。
79. 正因为没有空气，所以星星在太空中不像地球上看它们时那样闪闪发亮。
80. 日蚀和月蚀在世界各地并不是常常都能看到。
81. 既没有绝对的绝缘体，也没有绝对的导体。
82. 与固体不同，气体从来没有固定的体积。
83. 温度的绝对零度是永远不会达到的。
84. IBM 可能永远不能恢复它在计算机工业中无可争辩的巨人地位。
85. 马力和马毫不相干。

86. 一般情况下，淬硬钢材及玻璃和陶瓷等脆性材料，通常不宜使用金刚石切削。
87. 与固体不同，气体从来没有固定的体积。
88. 并非每一个人都相信比萨斜塔真的能够免于坍塌。
89. 不是所有的东西在刚被发现时都有用。

第4章

1. 复杂的计算机系统易于受软件故障所害。
2. 冰浮在水面上，因为它的密度比水小。
3. 在很大程度上，这是从这类系统中获得高精度的原因。
4. 研究和发展对于航空航天工业是特别重要的。
5. 使用润滑油是为了确保机器能正常地运转。
6. 万维网是一个独特的通讯和出版媒体。
7. 将地热能转变成电能是输送这种能量的切实可行的办法。
8. 滚针轴承的承载能量大。
9. 齿轮的齿必须具有足够的强度来运转。
10. 科学家们已经找到了一种适合于大多数岛屿的近海和沿海地区的测量方法。
11. 负载的电阻对于放大器来说太高。
12. 荒漠上稀疏地覆盖着灌木丛和杂草。
13. 在正常情况下，原子的整个质量都集中在原子核里。
14. 滚动轴承与滑动轴承相比最主要的优点在于它几乎完全消除了摩擦。
15. 为了检测各种气体，可以对空气样本进行化学分析。
16. 如果可能的话，这些盐应该用电来加热。
17. 金属不像塑料制品那样容易变形。
18. 这种材料在交变载荷下可以承受较高的应力。
19. 截面越大，强度越大。
20. 氢是最轻的气体，其密度为 0.0894 g/l。
21. 许多国家此时停止投资。
22. 这些机械近年来尤其是在美国应用得非常广泛和成功。
23. 在液体表面有表面张力。
24. 金子的颜色和黄铜相似。
25. 有许多物质，电流根本无法通过。
26. 项目费用不是个小数目。
27. 电缆沟旁设置的栅栏太矮了。
28. 很早就发现了人们用来制造机器的金属。
29. 高温使我们留在室外。
30. 全世界采用同样的数学式号和符号。
31. 由于机器噪音太大，我的精力难以集中在工作上。
32. 除了有点满载，轮船航行得很好。
33. 这台机器已经连续运转七八个小时了。
34. 这表明可能出现了意料之外的情况。
35. 当用瓦作为电功率单位太小时，我们可用千瓦。
36. 有时必须钻到 1,000 米或更深的深度，才能找到足够量的水。
37. 短波发送出去，遇到障碍就反射回来。

38. 由于这种电流不断地改变方向，故称为交流电。

39. 一些公司定期更换某些零部件，先确定零部件的有效寿命，以便在零部件即将磨损之前予以更换。

40. 这就是在工业上广泛使用表面活性剂的原因。

41. 氧的化学性质活泼，能参与许多反应。

42. 大规模的集成电路的发展将使现代化电子计算机以及其他装置越来越小。

43. “适当选择”这两个词应当予以充分强调。

44. 该机床是供磨削硬质合金刀片上下表面用的。

45. 聚合物是一种高分子的物质。

46. 这些机器每个月可以加一次或两次油。

47. 化学反应的速度与反应物的浓度成正比。

48. 光年是计量距离的最大单位。

49. 任何物质，无论是固体、液体或气体，都是由原子构成的。

50. 电压降低了4/5（电压为原来的1/5；电压降低到1/5）。

51. C是D的1/2。

52. 与旧式机器相比的主要优点是体积缩小了2/3（缩小到1/3）。

53. 第一部类的资本已经由6000增加到6500，即增加1/12。

54. 这种薄膜的厚度是普通纸的1/3。

55. 地塞米松的剂量是强的松的1/7。

56. 药品价格与1950年相比降低了75%。

57. 一码等于三英尺。

58. 这部新压缩机比旧的轻一半。

59. 把数据传输速度减少一半，将会使每一符号间隔时间延长一倍。

第5章

1. 如果把产生应变的力去掉，具有弹性的材料就会恢复到它原来的体积和形状。

2. 在空气中冷却之后，这种钢就变得越来越硬。

3. 将金属线很好地绝缘后，就可以用做导线。

4. 对于同样粗细的电线来说，铜比铝的电阻小，但由于铝的重量很轻，所以它单位重量的电阻较小。

5. 热的膨胀着的气体转动涡轮的叶片，因而放出了大部分的动力，驱动了压缩机。

6. 某些合金元素可精化钢的晶体度，从而提高了钢的硬度和强度。

7. 光子虽然没有物质质量，但可以认为它具有辐射质量。

8. 纯铁是银白色的金属，在1535 °C融化。

9. 各种分子的大小和重量都有很大的差别，包括从最小的微分子到最大的宏观分子。

10. 按其传导性来看，这种金属可能是铝。

11. 价电子能够脱离其原子，给原子留下一个空穴。

12. 静止的物体不具有动能，因为它的速度为零。

13. 电流把水分解为氢和氧，氧气从阴极释放出来。

14. 电能可储存在由一绝缘介质隔开的两块金属极板内。

15. 黑色金属是一种主要由铁构成的金属。

16. 就最后提到的方法而言，现有水头提供的水通过射流泵在其扩散管处形成真空。

17. 已知电压和电流，根据欧姆定律就可求出电阻。

18. 如果压力不变，一定量的气体的体积就与绝对温度成正比。

19. 与光学显微镜相比，电子显微镜具有极高的分辨率。

20. 煤和石油里的能量来自太阳，是由数百万年前的植物储藏起来的。

21. 有92个电子围着又密集又复杂的原子核旋转的原子，就是元素铀。它似乎是天然原子中最复杂的原子了。

22. 在建立疲劳破坏准则时，迟滞能量是一个有用的依据。

23. 近200年来，科学家、数学家和工程师们发展了结构理论，其目的在于通过计算给结构物的设计工作提供可靠的基础。

24. 当基极接地时，晶体管Q4便成为一个非常高的阻抗。

25. 作为阻抗函数的另外一种选择，可以采用摩擦系数查询表来描述小区之间的阻抗。

26. 这些关系式展示了核子结构的某些特征。

27. 必须保护感应电动机，防止过载和欠压。

28. 多孔壁的作用就像一把筛子，它把不同质量的分子分开。

29. 传送电磁波需要能量。

30. 为了保证反应进行得相当快，要求在少量液相和多得多的固相之间有最密切的接触。

31. 对这个量要精确地下定义是相当困难的。

32. 必须通过一个附加接地炭刷把转子大轴接地。

33. 当往熔融生铁中吹入氧气时，硅首先开始氧化。

34. 用并联电路使我们有可能开关任何电灯或电器而不影响到别的电路。

35. 绝缘体使导线相互隔离并且不接触到其他物体。

36. 热从高温流向低温的这一自然趋势使得热力机可以把热变为功。

37. 为了让硅晶体导电，我们必须寻找一种方法让一些电子在晶体内移动，尽管原子之间有共价键。

38. 单台发电机的容量越来越大，目的就是满足不断增长的用电需求。

39. 虽然只发现了107种元素，但是这些元素以不同的方式化合在一起就会形成世界上无数的千差万别的东西。

40. 在通讯系统中，电子学要解决的问题是如何把信息从一个地方传递到另一个地方。

41. 短路会使得很大的电流流过。

42. 动力使机器运转。

43. 摩擦迫使运动物体停止。

44. 直到19世纪证明了磁是电产生的时候，人们才开始对永久磁性的本质作认真的研究。

45. 绝缘体用来把电流限制在所要求的路线中。

46. 最小的电力系统如图1所示，它由能源、原动机、发电机和负载构成。

47. 电流通过导线时产生热这一事实是众所周知的。

48. 但是电能以光速——大约每秒186,000英里传播。

49. 从石油中提取合成蛋白以制作人造食品已被列为某些国家石油公司的当前首要科研项目。

50. 电工学主要关注的是通过输送能量做功的情况。

51. 在这类设备上，有可能将雷诺数提高3~9倍，而不用进一步提高速位差与驱动功率。

52. 磁铁具有与电荷相似的特性：同磁极相斥，异磁极相吸。

53. 没有足以克服阻力的力，静止的物体决不会运用。

54. 在这门学科的发展早期曾经假设过：电流是一种流体，或确切地说是两股流体，这两股流体等量地存在于物体中，同时还假设，给物体充电就是给该物体加过量的正电流或负电流。

55. 工件将通过一系列轧轮，轧轮的间隔依次缩小。

56. 金属超声波探测，就是把超声振动施于弹性金属材料，并观察振动在材料中产生的作用。

57. 额定电流被认为是由击穿模式决定的脉冲电流（峰值电流）和连续电流组成。

58. 它们一般会被分成反射波列和透射波列，这两者的相对强度取决于界面上速度变化的大小、变化地缓急度以及入射角的大小。

59. 表面磨损或者是由滑动面的热应力引起，或者是由滑动面的疲劳损坏造成的。

60. 业已证明，液体中的水既会影响液体的完全性又会影响系统元件的工作特性。

第6章

1. 降压变压器能把电压降低到所需要的任何数值。
2. 必须注意，以防电子仪器损坏。
3. 本文的目的是讨论核能的应用。
4. 压力随着深度而增加，这使人类进入深水很困难。
5. 某些工作条件可能会使机壳的温度过高，从而引起火灾。
6. 只有大脑处于工作状态，人体的各个部分才能正常工作。
7. 原子太小了，即便用高倍显微镜也看不见。
8. 我们发现，几乎物质所有的物理属性都受热的影响。
9. 这项研究发现，鱼在饮食中所占的比重与心脏病的死亡率有着直接的关系。
10. 这一现象表明，冰的密度比水的小。
11. 我们可以用两个串联的电阻，或者我们可增加现有的一个电阻的电阻值。
12. 交流电没有固定的方向，并且大小也在变化。
13. 线圈的各线匝之间以及与铁心之间均应绝缘，否则，将产生短路而不能形成流经线圈的电流。
14. 逆时针旋转的频率为正，顺时针旋转的频率为负。
15. 假如将食盐放在放大镜下观察，你就会发现每粒食盐都是一个有六个面的立方体。
16. 没有绝对的绝缘体也没有绝对的导体，因为一切物体对电流都有阻力。
17. 最先使用的材料是铜，因为铜易于炼取。
18. 分子的平均动能增加了，因而温度也升高了。
19. 质子数与电子数相等，因而整个原子呈电中性。
20. 位移是一种矢量，而时间是一种标量。
21. 导体越长，电阻越大。
22. 热机中的能量损失尽管很大，但是与能力守恒定律丝毫不矛盾。
23. 使用哪种润滑剂主要依轴承的转速而定。
24. 核武器毕竟是人类制造的，一定会被人类所销毁。
25. 当电通过灯泡里的细钨丝时，会使钨丝达到很高的温度。
26. 某些原子由于其结构上的原因容易失去电子。
27. 这种冒险的代价，不管是在人力还是在能源消耗方面，都将是巨大的。然而，许多人认为看，冰山牵引最终会证明比选择海水脱盐法花费要少。
28. 由于温暖的洋流能把温暖的气候带给那些本来十分寒冷的地区并使之变暖，因此，海洋在使我们这个地球更适合人类居住方面也扮演一个重要的角色。
29. 之所以讨论这个题目，是因为人们普遍认为，1976年高雷诺数研究专题研讨会落实的东西太少。
30. 用电解法生产的氢气几乎是完全纯的，虽然比热裂法生产的要昂贵一些。
31. 能量能从一种形式转换为另一种形式，所以电可以转变为热能、机械能、光能等等。
32. 当电荷运动时，就做功。
33. 随着范围的扩大，电波变得愈弱。
34. 水星背着太阳的一面终年黑暗，那里的温度只比绝对零度高几度。

35. 如果一种钢不仅含有碳、硫、磷，还含有1% 以上的锰或0.3% 以上的硅或一些碳素钢中不包含的其他元素，那么这种钢便是“合金钢”。

36. 控制系统可能包括一位派守在电厂的值班员，该值班员观察发电机输出终端的一整套仪表，并做一些必要的手动调整工作。

37. 大动脉和静脉不允许任何东西通过，然而最小的血管毛细血管，则可以让水分和其他微分子通过并进入组织。

38. 由于风把砂粒刮起来，碰撞大岩石，久而久之，较松软的岩石层就被慢慢地磨损。

39. 从经济观点看，集成电路的成本之所以大大降低，是因为采用了自动化批量生产方法。

40. 红外系统既可用于对空作战，也可用于对舰作战，但是它有两个固有的缺点：一是对光学能见度有一定的依赖性，二是易受天然或人工干扰源的干扰，因此，该系统的对空作用较之对舰的作用更为诱人。

41. 由上述可知，太阳的热量可以穿过太阳与地球大气之间的真空，而大多热量在通过大气层时都扩散和消耗了。实际发生的情况正是如此。但是热量的损失究竟达到什么程度，目前尚未弄清。

42. 我们从经由这个走廊进来的人身上已缴获的文件，加上近几个月从战俘那里得到的口供，使我们相信，渗透的规模扩大了很多。

43. 尽管中国的通讯落后，或者说正是由于这种落后，中国希望跳跃式地进入通讯数字显示时代，一下子越过耗资巨大的过渡技术阶段。这些过渡技术，工业化国家正在寻求用先进的数字显示系统取而代之。

44. 他把一个带有切口的齿轮装在钟上，再把它与电报线连接起来。这样安装好后，每次到点时，正确的电码点数就靠轮子的转动自动地沿电线发出了。

45. 如果做父母的对这种青少年期的反应有所准备，而且意识到这是一个标志，表明孩子正在成长、正在发展宝贵的观察力和独立的判断力，那么他们就不会感到如此伤心，所以也就不会因为愤恨和反对，迫使孩子产生反抗的心理。

46. 如果我们采取行动以便能够继续与中东问题各方始终保持接触，那么我们美国就能有效地担当起总统所提出的两项任务，那就是在中东结束敌对行动以及对该地区的永久和平做出贡献。这就是我们的观点。

47. 第二个方面是全体社会成员（从政府官员到普通公民）都使用科学家们在工作中所采用的特殊的思维方法和行为方法。

48. 总部设在深圳的一家中国内地的电子公司与香港政府签订一项合同，参与建设已规划完毕总投资为三亿一千两百万美元的一个第一流的半导体工厂。这家公司由此主动开始向香港发展迟缓的高科技领域注入几十亿美元的投资。此举体现香港与内地角色的互换，具有潜在的重要性。

49. 如果要用通常的方法来数银河系的星星，那么恐怕一辈子也数不清。

50. 月球上完全没有水和氧，一次月球是一个绝无生命的死寂世界。

51. 计算表明，对坦克的击毁概率介于30%和40%之间。

52. 科学家们面临着比以往任何时候都更为明确的任务，即把科技成果造福于人类。

53. 人们仍然相信，普通的人要比自然的力量或人类造出来的机器更伟大，而且最终会控制它们。

54. 坚持一个中国的原则是实现和平统一的基础和前提。

第7章

1. 于是，电被看作能量的一种形式，可把动力以电的形式迅速地，而且通常是非常高效率地由一个地方输送到另一个地方。

2. 此项过程与发送端所进行的相类似，载波被压制在平衡调制器内。

3. 在某些情况下必须给病人吸纯氧气。

4. 就功率来说，汽油发动机不算重，因此，现在的小型飞机仍然使用它。
5. 人类日益关注气候变化，关注节约能源和城市可持续发展，这推动了玻璃技术不断创新。
6. 候鸟的旅行情况，很多是靠在鸟腿上套个铝环作标记查明的。
7. 管子常被空气中的氧腐蚀。
8. 适当的润滑也可以大大减少摩擦。
9. 要寻求人类太空探测的合作途径，就必须应对一些至关重要的问题。
10. 后一项决定在2004年发布的预算案和新太空计划中就有所暗示，并且从未明确取消。
11. 目前，气象学家们正在努力确定这个大陆上的降水量中有多少是以这种方式产生的。
12. 人们曾一度认为，海洋的深处不可能有生命，因为那里没有光线，而且极冷。
13. 假定辐射损耗为5%，试计算所需的蒸汽量。
14. 相对于以往，他们坚信在新太空计划的参与权上将会拥有更多讨价还价的砝码。
15. 实际上空穴似乎是随外加电场的作用而运动，好像它是带正电荷和具有正质量的粒子那样。
16. 由于国际空间站正处于最后的装配阶段，它的运行机制已经在起作用。
17. 当然只有到达月球之后，月表探测系统才有意义。

第8章

1. 防御机制的系统发育也反映了特异免疫应答用来加强自然免疫这个概念。

2. 台风和飓风并不会形成特大的巨浪，因为这些风虽然很强，但它们并不能沿着同一方向长时间的吹。

3. 从图中可清楚看出高层次管理人员需要与设备性能有关的过去和未来的信息。基层修理车间的人员需要同样的信息。

4. 只要弹性圈的压缩性没有消失或受到任何影响，压力脉动对于接头的密封作用和可靠性也就几乎没有影响。

5. 太阳的体积约为地球的130万倍。

6. 科学定律常用文字来表达，但更多的是用公式来表示。

7. 任何物质，不管它是固体、液体或气体，都是由原子组成的。

8. 按传统说法，化学已逐渐发展成了四大分支：有机化学，无机化学，物理化学和分析化学。

9. 世界无需担心将来可能出现煤、石油、天然气或其他燃料来源短缺的问题。

10. 自动化要求不断地详细了解机器的操作，一有必要可以立即采取最佳校正措施。从这一确切含义来看，自动化只有在充分利用了它的三个主要因素，即通讯、计算和控制之后，才能完全实现。因此，我认为很有必要使大家认识和了解这种通讯、计算和控制三结合对我们社会的意义，至少是某些方面的意义。

11. 一氧化碳是一种无色、无臭、无味的气体，对植物、可见物和各种物品都没有害处。可是，由于它与红血球里的血红蛋白的结合加强了氧气与血红蛋白的结合力（见反应式3.1），就会造成导致人体和动物中毒的严重问题。

12. 在亚里士多德时代之前，人们记录光点在夜空中运行的轨迹，就已有好几个世纪了。

13. 人口增长的压力造成城市向高空发展，从而人为地提高地价，其结果是人们被迫适应城市的拥挤状况，住进了高楼。

14. 各种已知粒子所带的电荷或恰好等于电子电荷，或是电子电荷的整倍数。

15. 水向下所施加的压力和水的深度成正比。

第9章

1. 上午10点到下午5点

2. 1968 ~ 1972 年

3. 1944 年末到 1945 年初的那个冬季

4. 地中海中的岛屿和加那利群岛

5. 当时值得一提的有两件大事：一个是埃及工人对金属、陶瓷和染料的实际知识，另一个是对早期的希腊哲学家等人的研究。

6. 一些希腊的哲学家们认为一切物质都是由四种同样的元素构成，那就是土、火、空气和水。

7. 如果转基因动物具有繁殖能力，植入的异体基因（即转基因）就会遗传给下一代。

8. 我们将拥有人脑的路线图。这是 21 世纪的一个真正的尖端领域：人脑是我们所知道的最复杂的系统。它含有 1000 多亿个神经元（大概是银河系中星星的数量），其中每个神经元又连接到另外 1000 多个其他神经元。下一个世纪的早期，我们将应用先进的磁共振呈像形式绘制详细的人脑神经元运作图。

9. 用机械加工方法，特别是用磨削方法，可以获得最佳表面粗糙度。

10. 最轻的元素氢只有一个电子。

11. 尽管电子计算机有许多优点，但它们不能进行创造性的工作，也不能代替人。

12. 这所大学现有计算机科学、高能物理、激光、地球物理、遥感技术、遗传工程六个专业。

13. 伦敦《泰晤士报》的高级记者史密斯先生。

14. 在 1872 年，他用“电流的链、数学化处理”为标题在报纸上公布了这个结果。

15. ……大型卡车拖着装载出口木材的双层货架挂车，驶过马格德林和大学教堂。

16. 在七年内，将省、地、县、区、乡的各种必要的道路按规格修好（其中有些是公路，有些是大路，有些是小路）。

17. 1998 年 10 月，《时代》杂志列出了世界上最有影响的 50 位电脑时空研究专家，其中有两个中国人，一个就是负责中国政府建立的数据通讯网络、被这家杂志称为“中国英特网之父”的刘云杰。

18. 这里的差别足以证明：在决定何种管理训练对工长最有用之前，人们必须从工长所在的位置，他需要促动什么人和他有哪些进行促动的机会等方面先对他的工作进行一番考察。

19. 不锈钢硬度大，强度高。

20. 为便于学习起见，通常将这门学科分为力学、热学、光学、电学和声学。

附录　科技术语构词中常见的词缀

1. 前缀

(1) a-　非，不　asymmetry　非对称性；astray　偏离

(2) aero-　空气，飞机　aerodynamics　空气动力学；aerocraft carrier　航空母舰

(3) all-　全　all-weather transportation　全天候运输；all-round champion　全能冠军

(4) anti-　反，逆，抗，防，耐　antifreeze pump　防冻泵；antiphase　反相；antidote　解毒药

(5) astro-　天文，宇宙　astronavigation　天文导航；astrovehicle　宇宙飞行器

(6) auto-　自动，自　autocoder　自动编码器；autochart　自动流程图

(7) bi-　双，重　bipolar relay　双极继电器；bicarbonate　重碳酸盐

(8) bio-　生物，生　bionics　仿生学；biocatalyst　生物催化剂

(9) cine-　电影　cinefilm　电影胶片；cinemicroscopy　电影显微术

(10) co-　共，通　coaxial cable　同轴电缆；cosine　余弦

(11) cotro-　反，逆　controsurge winding　防冲屏蔽绕组；controflow　逆流

(12) counter-　逆，反　counteractant　中和剂；counterpart　对方

(13) de-　除，脱　defroster　除霜剂；derailment　脱轨

(14) deci-　十分之一，分　decimal　十进制；decibel 分贝

(15) di-　双，偶，两　diode　二极管；dibit　两位

(16) du-　二，双　duplicate　一式两份；duoplasmatron　双等离子体发射器

(17) equi-　同等，均　equipartition　均分；equilibrium　均衡

(18) ferro-　铁，钢　ferroalloy　铁合金；ferroxyl indicator　铁锈指示器

(19) gyro-　脱落，旋转　gyroscope　陀螺仪；gyrosphere　回转球

(20) hector-　百　hectare　公顷；hectoliter　百升

(21) hemi-　半　hemisphere　半球；hemipyramid　半椎体

(22) hepta-　七　heptagon　七角形；heptad　七价原子

(23) hexa-　六　hexagon　六角形；hexaploid　六倍体

(24) homo-　同，相似　homochromic　同色构异体；homoclime　相同气候

(25) hydro-　水，氢化　hydroelectric　locomotive　液力传动内燃机车；hydrofining　氢化提纯

(26) hyper-　高，超，重，过　hypertension　高血压；hypervelocity　超高速

(27) infra-　下，亚，次　infrared rays　红外线；infrastructure 地基

(28) inter-　相互，际间　internet　互联网；intercity train　城际列车

(29) iso-　等，同　isotherm　等温线；isotope　同位素

(30) macro-　大，宏观，常量　macrocontrol　宏观调控；macroanalysisi　常量分析

(31) mal-　不，失　malnutrition　营养不良；malfunction 失灵

(32) mega-　兆，百万　megastructure　特级大厦；megaton　百万吨级

(33) micro-　微观　microelectronics　微电子学；microfilm　缩微胶卷

(34) mono-　单，一　monoblock prestressed concrete sleeper　整体预应力混凝土轨枕

(35) multi-　多　multi-frequency locomotive　多频电力机车

（36）ortho- 正，直 orthocenter 垂心；orthograph 正投影图
（37）over- 过，超 overrun 超限；overexposure 曝光过度
（38）photo- 光电，光敏 photosynthesis 光和作用；photoreceptor 感光器
（39）poly- 多，聚 polymer 聚合物；ploysulphide 聚硫化物
（40）post- 后 pose-acceleration 偏转后加速；postmeridian 午后
（41）pseudo- 伪，假 pseudo-program 伪程序；pseudophotoesthesia 光幻觉
（42）quadr- 四，二次 quadruplicate 一式四份；quadraphonics 四声道立体声
（43）quasi- 准，拟 quasi-high-speed 准高速；quasi-ordering 拟序
（44）simi- 半 semi-final 半决赛；semitrailer 半挂车
（45）stereo- 立体的 stereooptics 立体光学；stereophone 立体声耳机
（46）sub- 子，亚 subloop 子回路；subsoil 亚土层
（47）super- 超 super-streamlined train 高级流线型列车
（48）tele- 远，电 telemetry 遥测术；telecommunication 电信
（49）thermo- 热 thermostat 恒温器；thermoscreen 隔热屏
（50）trans- 超，越，变换 transparency 透明度；transpassivation 过钝化
（51）tri-，ter- 三 tertiary 第三的；triplicate 一式三份
（52）ultra- 超，过 ultrasonic rail detector 超音波钢轨探测器

2. 后缀

（1）-er（or）器，机 air-oil booster 气—液增压器；air compressor 空气压缩机
（2）-ite 石，矿物 granite 花岗岩；sulphite 亚硫酸盐
（3）-meter 表，计 speedmeter 速度计；ohmmeter 电阻表
（4）-ity 性，度 absolute viscosity 绝对粘度；track elasticity 轨道弹性
（5）-scope 仪，镜 spectroscope 分光镜；telescope 望远镜
（6）-graph 仪 spectropraph 摄谱仪

3. 中缀

一般来说，中缀没有具体的意义，只是为了发音的便利而增加的一种字符，在科技英语中，中缀主要由元音字母充当，如：

“o” tachometer 转速计
“i” colorimetry 比色试验；technicolor 彩色印片法

参考文献

1 Eugene A Nida, Charles R Taber. The Theory and Practice of Translation. E J Brill, Leidene, 1982
2 Halliday. An Introduction to Functional Grammar. London: Arnold. 1985/1994
3 Halliday, Hasan. Cohesion in English. London: Longman, 1976
4 Peter Newmark. Approaches to Translation. Pergamon Press Ltd., 1982
5 Van Dijk. Text and Context. Cambridge: CUP, 1997
6 Van Dijk. Handbooks of Discourse Analysis. Cambridge: CUP, 1985
7 陈定安．英汉比较与翻译．北京：中国对外翻译出版公司，1998
8 戴文进．科技英语翻译理论与技巧．上海：上海外语教育出版社，2003
9 丁国声，周国培．新世纪理工科英语教程写作与翻译指导．上海：上海外语教育出版社，2004
10 范武邱．实用科技英语翻译讲评．北京：外文出版社，2001
11 冯志杰．汉英科技翻译指要．北京：中国对外翻译出版公司，2001
12 冯庆华．实用翻译教程．上海：外语教育出版社，2002
13 方梦之．科技英语实用文体．上海：上海翻译出版公司，1989
14 耿红敏．实用英汉翻译．上海：复旦大学出版社，2005
15 胡壮麟．语篇的衔接与连贯．上海：上海外语教育出版社，1994
16 黄国文．语篇分析概要．长沙：湖南教育出版社，1988
17 李惠林．上海科技翻译．上海：上海翻译出版公司，2003
18 秦荻辉．科技英语写作．北京：外语教学与研究出版社，2007
19 汪淑钧．科技英语翻译入门．广州：广东人民出版社，1979
20 王德军等．实用英汉翻译教程．北京：国防工业出版社，2007
21 夏喜玲．科技英语翻译技法．郑州：河南人民出版社，2007
22 杨士焯．英汉翻译教程．北京：北京大学出版社，2006
23 严俊仁．科技英语翻译技巧．北京：国防工业出版社，2000
24 严俊仁．英汉科技翻译新说．北京：国防工业出版社，2010
25 张培基等．英汉翻译教程．上海：上海外语教育出版社，1980
26 朱永生等．英汉语篇衔接手段对比研究．上海：上海外语教育出版社，2001
27 赵玉闪等．科技英语翻译．北京：中国计量出版社，2008
28 赵萱，郑仰成．科技英语翻译．北京：外语教学与研究出版社，2006
29 赵萱，郑仰成．科技英语翻译教师手册．北京：外语教学与研究出版社，2006